COMPUTER-AIDED CHEMICAL ENGINEERING, 1

DISTILLATION DESIGN IN PRACTICE

L.M. ROSE

Privatdozent, Technisch-Chemisches Laboratorium,
Eidgenössische Technische Hochschule, Zürich, Switzerland

ELSEVIER
Amsterdam — Oxford — New York — Tokyo   1985

CHEMISTRY 7302-8903

ELSEVIER SCIENCE PUBLISHERS B.V.
Molenwerf 1
P.O. Box 211, 1000 AE Amsterdam, The Netherlands

Distributors for the United States and Canada:

ELSEVIER SCIENCE PUBLISHING COMPANY INC.
52, Vanderbilt Avenue
New York, NY 10017, U.S.A.

Library of Congress Cataloging-in-Publication Data

Rose, L. M., 1935–
    Distillation design in practice.

    (Computer-aided chemical engineering ; 1)
    Includes bibliographies and index.
    1. Distillation apparatus--Design and construction.
    2. Computer-aided design. I. Title. II. Series.
    TP159.D5R67 1985   660.2'8425   85-16845
    ISBN 0-444-42477-6

ISBN 0-444-42477-6 (Vol. 1, hardcover)
ISBN 0-444-42481-4 (Vol. 1, paperback)
ISBN 0-444-42482-2 (Series)

Printed in The Netherlands

PREFACE

Over the last 20 years the computer has gained increasing importance in design, particularly in distillation design. The stage has now been reached where all industrial distillation design of any consequence is done using computer methods.

This should have some effect on teaching. The same fundamental principles still hold, and they must still be taught, but the students should also be introduced to computer methods; shown the principles behind them and shown how to use them more effectively and intelligently.

This text is designed as a senior course in distillation for the last undergraduate year, or a postgraduate course, to link the theory of distillation given in the standard courses to industrial practice. It makes a useful senior course in design also because it integrates various disciplines - thermodynamics, design and control, as well as distillation itself.

The course was developed as a senior distillation course with industrial attendees for Lappeenranta Technical University, Finland, and has since been held a number of times at the Swiss Federal Technical University (ETH), Zurich. Run as a workshop with 50% of the course time being computer exercises, the course is received very enthusiastically as long as the attendees already have had a grounding in distillation. The computer exercises are most appreciated by industrial attendees. Such exercises are included in the Appendix.

If we are to teach design properly in an up-to-date course that makes the student familiar with using the computer for engineering design, then it is essential that the students are allowed to use design programs during the course. This requires that suitable teaching software be available. For this course EURECHA (European Committee for Computers in Chemical Engineering Education) software was used. The set of programs, DISTILSET, was the source of the binary and multicomponent programs. Data bank and flash calculations, together with the supply of VLE data for the distillation programs was supplied by the EURECHA data bank CHEMCO. The batch distillation exercise used the EURECHA BATCH program, and exercises in column dimensioning and internals selection can be investigated using the EURECHA program INTERN.

The manuals for these programs are given in the Appendix, where information on where the program can be obtained is also given.

I would like to record my thanks to Professor Nystrom and Palosaari, who arranged for me to give the course in Lappeenranta, and so gave me justifiable reason to get the material together in an orderly fashion.

Thanks are also due to Elsevier, who have agreed to publish a student paperback edition in order to bring the book within the reach of students; I only hope that sales will justify this decision.

For the production of the book I would like to thank Ingrid Rasmuson for the typing and Kurt Eigenheer for his work on the figures.

And lastly, as always, I wish to thank my family for allowing me my long periods of absence and inattention.

Zurich 1985   L.M.Rose

CONTENTS

Nomenclature

A	constant in Antoine equation	
a	liquid phase activity	
a	constant in Redlich-Kwong equation of state	
a	constant in BWR equation of state	
A_o	constant in BWR equation of state	
a	surface area per unit volume	(m^{-1})
B	constant in Antoine equation	
B	second Virial coefficient	
b	constant in Redlich-Kwong equation of state	
b	constant in BWR equation of state	
B_o	constant in BWR equation of state	
B	Bottoms flow	$(kmol\ h^{-1})$
B	Boiler content	$(kmol)$
C	constant in Antoine equation	
C	third Virial coefficient	
C_o	constant in BWR equation of state	
C_B	equipment item base capital cost	1000 $ (1968)
D	distillate flow	$(kmol/h)$
D	molecular diffusivity	(m^2s^{-1})
D	diameter	(m)
E_{og}	Murphree gas point efficiency	
E_{MV}	Murphree vapour plate efficiency	
E_{ML}	Mulphree liquid plate efficiency	
f	fugacity	
F	feed flow	$(kmol/h)$
F_p	packing factor	(m^{-1})
F	column loading F factor	$(kg^{1/2}m^{-1/2}s^{-1})$
f_I	inflation factor for capital cost estimation	
f_M	material factor for capital cost estimation	
f_P	pressure factor for capital cost estimation	
f_t	type factor for capital cost estimation	
G	Gibbs free energy	
G^E	total Gibbs free energy	
G	mass gas flow rate	$(kg\ h^{-1})$
h	height	(m)
h	heat loss per plate	(kW)
H	molar enthalpy	$(kJ\ kmol^{-1})$
ΔH	heat of vaporisation	$(kJ\ kmol^{-1})$
H	hold-up	$(kmol)$

K	vaporisation K-value	
k	film mass transfer coefficient	$(kmol\ h^{-1}m^{-2})$
K	overall mass transfer coefficient	$(kmol\ h^{-1}m^{-2})$
L	liquid flowrate	$(kmol\ h^{-1})$
L	length	(m)
L'	mass liquid flow rate	$(kg\ h^{-1}\quad)$
m	slope of equilibrium line	
N	number of plates	
n	molar flow	$(kmol\ h^{-1})$
n	number of components	
p	partial pressure	(bar)
P	saturated vapour pressure of pure component	(bar)
Q	molecular surface area data for UNIFAC	
Q	heat flow or enthalpy change	$(kW\ or\ kJ\ mol^{-1})$
q	heat condition of feed	$(-)$
R	gas constant = 8.314	$(kJ\ mol^{-1}K^{-1})$
R	molecular volume data for UNIFAC	
R	reflux ratio	
T	temperature	(K)
t	time	(h)
t	thickness	(m)
V	volume of 1 kmol of gas	(m^3)
V	mole fraction vapour	
V	vapour flow rate	$(kmol\ h^{-1})$
WC	Wilson coefficients (energy)	$(kJ\ kmol^{-1})$
x	liquid mole fraction	
y	vapour mole fraction	
z	compressibility factor	

Greek

α	constant in BWR equation of state	
α	relative volatility	
γ	liquid activity coefficient	
Δ	difference	
ε	voidage	
ρ	density	$(kg\ m^{-3})$

λ	molecular interactions	$(kJ\ kmol^{-1})$
μ	viscosity	
μ	fugacity coefficient	
π	total system pressure	(bar)
σ	standard deviation	
ω	acentric factor	

Sub- and superscripts

bu	bubble
B	bottoms
c	critical
C	combinatorial (molecular size) contributions (UNIFAC)
D	distillate
f	final
F	feed
g	gas phase
HK	heavy key
i	component
L	liquid phase
LK	light key
n	plate number
o	initial or inlet
or	orifice
ow	over weir
r	reduced
R	residual (interaction) contribution (UNIFAC)
V	vapour
w	weir
'	denotes a working variable only

Plate numbers, and component numbers are denoted by subscripts.

Chapter 1

DESIGN

1.1 DESIGN OBJECTIVES

The primary objective of "design" in chemical engineering is to determine the size and shape of equipment - in our situation, distillation equipment - that will perform satisfactorily, together with the other parts of the plant to produce the required product at required quality at the required rate.

Satisfactory performance means that the total cost must be acceptably low, that the plant should operate safely with minimum environmental pollution and that it should be flexible and robust, i.e. it should still operate satisfactorily when variations appear in the plant feed rate due to upsets in upstream equipment or due to changes in raw material quality.

The low-cost requirement is dictated by the economic reality that if you do not operate your plant in the most economic way, a competitor has a chance of improving on your process and taking away your business by being able to offer lower prices. Hence, great efforts should be made when designing to find the design that offers the lowest cost - to 'optimise' the design as it is sometimes called. In reality this does not mean to locate exactly the minimal total cost point in the mathematical sense - neither information nor sufficient time is available for such an exhaustive search - but considerable efforts should be made to see that no lower cost designs have been overlooked through inadequate investigation of the design alternatives.

As a secondary objective we can require that the plant be sufficiently flexible and robust to make plant operation possible when the design feed condition is not held, because variations have occurred from upstream processes. This objective conflicts with the first objective of low cost, because all flexibility must be paid for in terms of increased size of equipment or additional piping arrangements. For this reason, considerations of flexibility should be limited to variations of input that are likely and feasible for the upstream process. There is no point in designing for fluctuations that may never occur.

The factors 'safety' and 'minimum effect on the environment' are very strict requirements, legally enforced, and so these usually move from being 'objectives' to being 'constraints' - i.e. unless the regulations are adhered to, the plant cannot be built.

A further objective of a good design is that the final equipment should be easy to operate and control. This again will be in conflict with the low cost objective, and it may not command much sympathy from the design engineer when the operating personnel plead for an extra storage tank in the process. But it is reasonable to require that a plant that will run for 20 years should not produce problems every week because of off-specification material or any other cause associated with a 'tight' design.

A good design is therefore far more than simply the specification of some low cost equipment. The different criteria of low cost, flexibility and operability often conflict; they cannot easily be reduced to a common denominator and we therefore have a multiple-criteria optimisation problem.

1.2 THE DESIGN PROBLEM

Design in the broadest sense is always concerned with selecting a proposal that
has a number of different attributes that cannot be brought together under a
common denominator. Systems involved are very complex, and the objectives are
multiple - as in the design of distillation equipment.

Probably the most advanced thinking on 'design' is done by our architect
friends, since their problem is an extreme example of a multiple-criteria
problem involving functionality, cost, durability and aesthetics. There is a
great deal of similarity between architectural and chemical engineering problems, as can be seen when one compares the design of buildings such as facto-
ries and hospitals with chemical plant. In such cases some of the layout
problems are identical to plant layout and the traffic flow in a hospital
can be compared to the material flow in a plant; both dictate the layout.
Unfortunately in chemical engineering design the aesthetic aspects, so important to the architect, are almost always ignored!

The architects have evolved a general 'design method' which has stood the test
of time; it is represented by the algorithm described in Figure 1.1 .

The first step is the definition of the objective or objectives, as main and
secondary objectives.   The second step is the proposal of a solution by whatever means is available. The engineer often has a multitude of design procedures to help him generate a solution. The architect has to generate solutions
by much more abstract means. The engineer also often has to develop solutions
in an abstract way before he can apply his design procedures; for example, he
must himself generate the process flowsheet or distillation sequence before he
can start using a design procedure.

The next step is to test the solution. If we are designing a teacup, then we
can actually make a cup accurately to our design and use it. A building or
a chemical plant cannot be tested so easily, and so this stage represents the
work-up of the idea in more detail, further drawings, consideration of resulting implications etc..

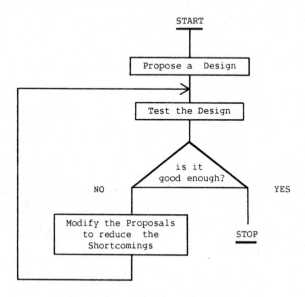

FIGURE 1.1 THE DESIGN METHOD

This detailed study will identify the weakness of the proposal, and a decision can be made as to whether the design is acceptable. Where this is not the case, a modified design is proposed, aimed at removing the weaknesses of the design already studied.

Note that the procedure is one of evolution. Systems are complex and so the best procedure is one of improving a suggestion until an acceptable solution emerges, rather than randomly generating designs until a good solution occurs.

1.3 CREATION OF ALTERNATIVES

How is the first proposal generated in order to carry out this design method? Presumably those with a spark of intuition, be they architects or engineers, derive their first proposals from their 'experience' and 'feel' for the problem. It is however useful to know that there are more systematic methods that can be applied by the less gifted of us, or by the gifted who want to check that they have not overlooked any useful possibilities.

Two essentially different methods exist; Critical Analysis and Morphological Analysis. In Critical Analysis, an existing proposal is analysed in great depth so firstly all the reasons and requirements for the proposed method are determined. Then all possible alternative ways of fulfilling the requirements are noted, using a systematic questioning procedure to try to cover as many possibilities as possible. The third stage is the subjective selection of the most promising possibilities and the working of them up into a possible solution, or possible solutions. The questioning procedures are as follows:

ROUND 1: To establish why the existing design is used. For every single operation in the total design, ask:
>   Why is it done?
>   Why is it done that way?
>   Why is it done at that time (or in that sequence)?

ROUND 2: To create alternatives for consideration. Every single operation studied above should be subjected to the following questions, and any alternatives (however wild) should be noted down.
>   Does it need to be done?
>   - don't do it!
>   - do it more intensively !
>   Do it in a different way
>   Do it before or do it later

ROUND 3: In this round, the alternative proposals for the single operations - which will have resulted in very many disconnected proposals - are considered together, and some promising possibilities for the complete system are developed. This stage requires intuition and judgement, but at least there has been a wide variety of possibilities presented before intuition is called upon.

The morphological analysis is a different approach, which ignores previous solutions and develops proposals in a combinatorial way by firstly defining all the objectives of the component parts of the total design. Then all possible solutions to satisfy these objectives for each subsystem are proposed. The design of the total system is then the combination of these subsystems to provide the complete range of solutions where every suggestion per subsystem is combined with every suggestion of every other subsystem. The result is of course an enormous number of possible designs, from which a design - or a few designs - are selected by intuition.

Intuition and experience always play an important role in design. The function of a design synthesis procedure is to derive as complete a set of possibilities as possible, to reduce the chance of the designer 'overlooking' a good solution.

1.4 SELECTION OF ALTERNATIVES

The Morphological and Critical Analysis both require alternatives to be selected from a wide range of possibilities. The design procedure shown in Figure
1.1 requires that the designs be tested against its various objectives for a
judgement to be made.

In the multi-objective situation the various objectives often conflict. How can
we make a selection or decide that a change has resulted in an overall improvement if we are not willing to leave this entirely to overall subjective
judgement.

1.4a A PRACTICAL SOLUTION

Being more specific, if we return to our distillation design, which has the 3
factors – cost, flexibility and operability – as objectives, how can we compare
and select alternatives? If the selection cannot be agreed between the operating personnel and the design team on the basis of 'sound engineering
judgement' then we require some procedure to enable the selection to be made.
For such a case, a simple decision making procedure does exist:-

The first step is to subjectively give each objective a weighting – say from 1
to 10. Ten would mean an extremely important objective, whereas one would be of
minor importance. Having weighted each objective, the next stage is to
subjectively judge each alternative for each objective and award points – again
between 1 and 10 – for the performance of the alternative with respect to each
objective. Ten would mean the alternative is excellent with respect to the objective in question; one would mean it performs very poorly.

We can now produce a table of the alternatives, objectives, and awarded points
and weights, as shown in Table 1.1, and obtain the weighted sum for each alternative.

Table 1.1 shows that alternative B should be chosen because it has the highest
weighted score. It is of course quite true that B is preferred because the
weighting on cost was as high as 10. The value 10 was as assigned as an independent subjective judgement without knowledge of the outcome, and therefore
should be worthy of more credibility in the decision-making compared to a
subjective choice of B without such a breakdown. Once the results are known,
revision of the weighting and scores is possible. Also bounds can be set on the
validity of the decision. In the case of Table 1.1, alternative B remains the
correct decision until the weights on the low-cost objective fall below 6. The
design team could be asked " Does anyone believe that the low-cost objective
weighting should be below 6?"

TABLE 1.1
SELECTION OF ALTERNATIVES IN A MULTI-OBJECTIVE SITUATION

	Objective 1 low cost	Objective 2 flexibility	Objective 3 operability	Weighted total
Relative weight	10	2	1	wt x score
Alternative A	7	10	5	95
Alternative B	10	2	1	105
Alternative C	5	10	10	80

This approach can be useful in arriving at a decision when no one is willing to accept or give an overall engineering judgement. It gets the project moving when it has halted at a decision point.

Although this method is still 'subjective' in that it is entirely based upon the subjective weightings and scores, it does break the problem down into component parts that may be more easily, or less preconceivedly, judged.

1.4b SOME MATHEMATICAL SOLUTIONS

Multiple-Criteria Decision Making

Operations Research is the science of the operation of large systems, and in Decision Analysis - one of its branches - it has long been recognised that there is a problem with regard to decisions involving real systems which have a number of objectives.

Rigorous mathematical treatment of such systems results in the production of 'Pareto Surfaces' in n dimensions, where n is equal to the number of objective functions. To remain with a geometric description of the method let us consider an example containing two objectives, both of which are quantifiable; the availability and capital cost of a distillation system.

Now, given a model of the distillation system that calculates availability and investment costs for any set of input variables defining the system (diameter, number of plates, reflux ratio, tank capacities, etc.), then this model can be optimised for one objective for a given value of the second objective. Hence a series of optima can determine the minimum plant cost if the availability to be used is 80%, 85%, 90%, 95%, etc.. These points can be joined together to form the 'Pareto Surface' for the system (see Figure 1.2).

The Pareto surface is the 'locus of efficient points'. Moving into the surface is sub-optimal for all objectives, moving out of the surface is infeasible, and moving along the surface causes an improvement of one of the objectives at the expense of others. Clear therefore, the chosen design should be a point on this surface. Exactly which point cannot be selected mathematically because it depends on the relative importance of the various objectives, and this cannot be defined quantitatively.

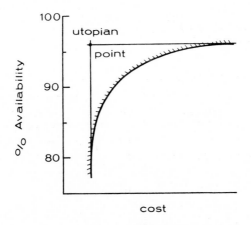

FIGURE 1.2
PARETO SURFACE FOR THE DISTILLATION SYSTEM.

Having presented this figure, the final decision is still left to the subjective judgement of the engineer, but now he has a clearer picture of the trade-offs involved. For instance, are the difference in costs between 80% and 90% availability so small that low values of availability are not worth considering? At the higher end, the cost of providing that last 1% availability is clearly prohibitive. The engineer can now select from the curve exactly which point he will accept for his design.

The method can be compared to the creation of a weighting with which to combine the different objective functions to make up a single quantitative objective function:

$$\text{e.g. Objective} = \text{wt x availability + Capital Cost} \qquad (1.1)$$

This is a much cruder procedure because the weighting 'wt' is difficult to select. Mathematically speaking, the two methods do not give the same results and the Pareto surface method gives the theoretically more defensible solution.

When the Pareto surface is developed for complex engineering systems containing standard equipment sizes and other non-continuous relationships that are part of engineering design, then the curve is far from simple, as shown by Figure 1.3. The variations in slope at different parts of the curve show that there are locally preferred points where relatively much is gained for one objective for relatively little loss in a second. The star on Figure 1.3 represents a relatively better choice that the circle.

Though the presentation of such information is useful to the engineer in making his choice, it can rarely be applied because of the computational effort involved. The creation of the surface requires sets of optimisations to be performed, leading to enormous computational times. However such methods are worthy of mention because they do rigorously define the multi-objective problem, and help give a better understanding of the designer's problem.

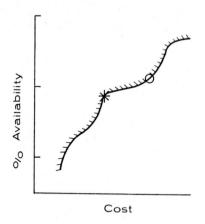

FIGURE 1.3
UNEVEN PARETO SURFACE ASSOCIATED WITH REAL SYSTEMS

Decisions Under Uncertainty

We have so far been discussing decision taking when the information, and hence the resulting objective function are known with certainty. It is however usual that the data used for the calculation of the objectives are uncertain. This uncertainty can be 'known' i.e. the standard deviation for the parameters of the VLE data can be calculated from experimental error - or 'unknown'- e.g. the yearly throughput, which will depend upon the future market demand.

When this is the case, then mathematical methods exist for computing the resulting uncertainty in the resulting objective function. The resulting distribution of the objective function is then useful in making a decision because the 'tail' of the distribution is an indication of the 'risk' associated with the decision. When selections are being made from alternatives, then the 'tail' of the distributions of all alternatives can be considered and the most risky may be discarded on these grounds alone.

The mathematical methods used for the creation of the distributions can be combinatorial, integration or Monte-Carlo methods. Further detail is not appropriate for a distillation text, but readers are urged to buy a text on Engineering Decision under Uncertainty. The subject is returned to in Chapter 3 in discussing Safety Factors in design.

The new mathematics of 'Fuzzy sets' deserves a mention in that it attempts to convert qualitative linquistic statements into a semi-quantitative form which can then be combined together with other quantitative statements to produce a semi-quantitative result.

Consider the following statements borne out of experience:-

   If the throughput is large, use large packing

   If the separation is difficult, use small packing

   If the feed is dirty , large packing is preferred.

Large, difficult, dirty, small and preferred are 'fuzzy' concepts. Each has an appropriate 'Grade of Membership Function' defined, which enables the conditional statements to be converted to numerical statements. Then, given any particular case - e.g. 1200 t/y, relative volatility of 3.1, and 2 g/m^3 solids - then following the conditional statements the Fuzzy mathematics will define the corresponding best choice of packing.

The method is a welcome attempt at developing techniques for methods that are not precise. The method as yet does not find application in chemical engineering design, but again, to understand the concepts gives a better understanding of the problems of design, which are often concerned with arriving at decisions which have qualitative aspects.

Finally, a brain child of Informatics is the Expert System. This is centered on the computer coding of all the statements that an expert is aware of when he makes a judgement. Together with these statements, probability distributions can be stored.

Hence for any question poised to the Expert System, the total set of rules are evaluated, and an answer (with probabilities) results. The methods were developed for medical diagnosis and geological predictions, but they could equally well be developed for the selection of column internals, for instance.

The systems are claimed to perform at least as well as the expert, and even better in that all the rules every time are systematically processed, leading to an improved reliability over the expert himself.

Notice that fuzzy sets is a form of mathematics, and Expert Systems is a form of information retrieval.

All these methods are as yet in their infancy and are in no way developed to the stage where they could be applied to everyday design. It is however interesting to be aware of them, because they highlight two interesting properties of the Design Decision - the multiple-objective and the subjective aspects. For practical distillation design however, we must await further developments of the methods and for the present stay with our subjective judgements and simple weighted analysis for our decision making.

1.5 COMPUTER-AIDED DESIGN

It is clear from the description of the design method that for complex systems there must be a great deal of trial and re-trial of new proposals. One would therefore expect that the easier it was to evaluate a new proposal, the more alternatives would be investigated and the better would be the resulting design.

In architecture, the new proposals usually involve the production of new drawings before an assessment can be made. In chemical engineering, assessments can generally be made after calculations have been performed which define the implications of the new design on cost and operating flexibility.

By introducing the computer in order to aid the design process, we are simply using the speed and convenience of the computer to make the assessment of the new design as easy as possible; ideally, one should have an almost immediate assessment to maintain a continuous train of thought during the design process. If used properly, such a tool should enable a better design to be achieved, both faster and cheaper.

Computer aided design (CAD) is a term employed for this activity, whether it be architecture, mechanical engineering or chemical engineering. Though the same term is used, the resulting use of the computer depends upon the application. In architecture, CAD refers almost exclusively to computer graphics and computer drafting. In mechanical engineering and electrical engineering the term is also usually associated with graphics. However, in chemical engineering the term applies to numerical computations associated with equipment design procedures.

In distillation design the advantages of CAD are numerous:

(a) A better final design should be achieved by the more thorough testing of alternatives and the testing of more alternatives.

(b) The computer result is accurate and consistent. The effect of small changes can be evaluated from the results obtained. In long, complex, manual calculations there is often a minor error or slightly different assumption that means that no reliable interpretation can be put on a small change in the results.

(c) Within a company, a company standard design method can be decided upon and this can be coded as the CAD software. Then all in-company designed distillation systems will have used the same method. Hence all designs will either be right, or all wrong in the same way. Any design errors thus become clear, and the standard company method and the design software are modified for the next generation of designs. In this way a company can gain confidence in its design method, which in turn leads to a reduction of the design safety margin and reduced costs.

(d) The engineer is usefully employed in making engineering judgements and decisions, and not in spending his time repetitively iterating through manual calculations.

All aspects of CAD in chemical engineering are not positive and unless CAD is applied in an appropriate manner, using well designed software, these negative aspects will be genuine drawbacks to CAD.

The most frequent heard criticism is that the engineer loses all control of his problem and uses the software supplied to him as if it were a 'black box'. "He puts figures in and receives an answer he must accept even though he doesn't know how it was arrived at. In the end, the engineer will forget all he ever knew about distillation and simply be a feeder of data to the program."

There is considerable truth in this comment, but since CAD is here to stay, we must devise methods of overcoming these problems. Much depends on education. If engineers are taught how programs work, they need never treat them as black boxes. So CAD must be properly introduced into the undergraduate curriculum. It is true that to carry out a calculation manually instils the principles of the design method into the engineer, and the calculation of the intermediate results (concentration and temperature profiles, gamma values, flooding velocities, weeping, etc) give the engineer a good 'feel' for his problem.

To counteract this loss in CAD , the appropriate software should print out all intermediate results that are of interest to the engineer. The distillation program that gives as the result no more than the number of plates and reflux is of little use in the design context because the engineering understanding necessary to make the next improvement is not there. It is again a matter of education to teach the engineer to learn as much from an output of a good program as he would learn from doing the calculation by hand himself.

Computer programs often require a good starting estimate before they can begin a calculation. This should be seen as an advantage in that it forces the engineer to make a 'back-of-an-envelope calculation' himself to decide on reasonable input data for the problem. Once he has done this, his understanding of the problem and appreciation of the results are much enhanced.

A further criticism of CAD is that problems are over-studied. As one design section manager once said "Twenty years ago we got the light-ends design out in two months without a computer, now with a computer it takes three months".

This frustration is brought about by the engineer carrying out the CAD not having been properly trained to study the problem only as deeply as is necessary to produce the design specification. It is a fascinating challenge to try a never-ending list of alternatives to get a finer and finer design, but this is way beyond the accuracy demanded for the design specification, and it should not be undertaken. Again, training should teach the engineer to bear this in mind.

1.6 SOME RULES FOR THE USE OF CAD

To avoid the criticism often levelled at CAD, the following rules should be observed in all CAD work:-

RULE 1 Start with an order-of-magnitude hand calculation, so that reasonable input data is given to the program and the results received are not a complete surprise.

RULE 2 Once the results of the first run are available, inspect them carefully and see that the whole program output is sensible and error-free before embarking on multiple case-study runs.

RULE 3 Investigate the importance of the various parameters in accordance with qualitative engineering expectations of the system, and not according to some empirical search procedure or experimental plan.

RULE 4 Note the savings achieved as the design is improved, and stop the work
when the further savings are no longer significant. Do not overstudy the problem.

RULE 5 Analysis of flexibility and operability are more important than an over-
extensive cost optimisation study (i.e. don't "fine-grind" the design, but
check other operating aspects).

1.7 SOFTWARE REQUIREMENTS FOR CAD

Before one can consider using CAD, one must of course have programs available.
It is appropriate here to make general comments on the specification for such
programs.

First and foremost the programs must be 'user' or more precisely 'engineer'
friendly. The engineer may use the programs perhaps once every three months.
He will never be an 'expert' with the program, and will never learn   and   remember special details of each program. For example one program requires names
to be 6 characters another 4, one uses density in g/cm^3 another kg/m^3!   To be
user-friendly to an intermittent user, all programs should be standardised, to
work with one philosophy, and to belong to one standard system.

1.7a INPUT

Interactive input is a user-friendly aspect that is useful, particularly for
small programs. For complex programs requiring much input, the engineer
has already studied the input requirement from the manual and prepared his
string of input data according to some specified table in the manual. In this
case he is quite happy to type in his string of data without waiting for the
endless list of politely worded 'prompts' to be printed out. Hence for larger
problems, facilities should also exist for inputting data strings independent
of prompts. Free format for such strings is a 'must' to prevent engineer frustration with incorrectly formatted input.

Input checking is important to prevent calculations being done with incorrect
data, both because of wasted computer time and the chance that a program error
will result which may cause unnecessary confusion and unjustified criticism of
the software.   In interactive mode, an input error should result in a repeated request for that data. In unprompted data-string input, an input error
must end in an error message and a termination of the calculation.

For large data sets , data once entered should be stored for later use so that
only corrections to the data set are necessary to run new cases.

1.7b OUTPUT

The output should be understandable to the occasional engineer-user.   It
should describe the progress of the calculation so that it is clear what the
program has done and what the value of the important intermediate variables
are. Some optional output control (e.g. short form, normal and debug
facilities) may be useful, but such control does bring dangers with it.   Our
RULE 2 requires careful inspection of the whole output for errors. If the user
has the option to reduce the output, or, as in some systems, to edit the output, then errors may not be spotted and problems only partially understood.

Printed outputs are essential for large problems. It is impossible to study a
distillation program output or a process flowsheet output properly on a CRT
screen.

1.7c   DATA

Physical property data must be available for the necessary computation in the
form of parameters to correlating equations and not as curves or tables in a
data book. All data should be in the same units. To have some Antoine
Constants in ln and bar and others in $\log_{10}$ and mmHg is a common occurrence,
but it is an absolute disaster for CAD. Of course, all engineers can convert
units, but they can also make mistakes on the way!

TABLE 1.2
EURECHA PROGRAMMING STANDARDS

These standards were developed to aid the interchange of teaching
programs between universities

- Standard 1966 ANSI FORTRAN
- Free-read input
- Common input philosophy
- Usable in both batch and interactive modes
- Incorporation of input checks where possible
- Pre-checking of log and sqrt statements to prevent
   program failure through negative argument
- SI units both in input and calculation
- Alphanumerics in A2 (integer) or A4 (real)
- Documented with user manual and loading instructions,
   other detail given as comments in listing
- No graphics
- Clearly identified file handling (not in 1966 standard)

These points concerning data logically point towards the use of a 'data bank'.
These are described in more detail in the next chapter, but it is fair to say
that the use of data banks - where all required physical property data are
consistently stored in a file which can be called up by any engineering program
requiring the data - is an important factor that will convert even the
most conservative engineer to CAD. The difference it makes to the effort required to run a distillation program is considerable. In fact, in teaching it
enables a course to thoroughly study distillation in a 2-3 hour workshop period. Without the data bank, it takes 2 - 3 hours alone to get the physical property data error-free into the program.

A physical property data bank is essential.

1.8 FURTHER READING

Rose, L.M., 1976. Engineering Investment Decisions - Planning under
                Uncertainty, Elsevier.

Zadeh, L.A.,1973. IEEE Systems, Man and Cybernetics, Vol SMC - 3, p28.
                - early outline of fuzzy sets.

Zeleny, M., 1982. Multiple Criteria Decision Making, McGraw-Hill.

1.9 QUESTIONS TO CHAPTER 1

1) What are the objectives, and what are the constraints of process engineering?

2) Why is it necessary to look at alternative designs?

3) What methods exist for choosing between alternatives?

4) How can the computer help in process engineering?

5) What are the dangers in using the computer for engineering design?

6) What general specifications should be defined for engineering software?

7) What role does the data bank play?

Chapter 2

VAPOUR/LIQUID EQUILIBRIA DATA

INTRODUCTION

Distillation is the separation of liquid mixtures of different components into a number of fractions of differing composition; often the objective is to separate the mixture into its pure components. This separation is achieved because there is usually a difference in composition between a liquid mixture and the vapour in equilibrium above it. Hence, by careful management of repeated vaporisation and condensation from a liquid mixture, pure component fractions can often be achieved.

Distillation is thus based on the difference in equilibrium compositions between vapour and liquid phases of a mixture. The characteristics and reason for the equilibrium should be understood. The ways in which these equilibrium data can be quantitatively recorded for later use in distillation design and possible ways of predicting such data are natural pre-requisites for any course on distillation design.

It will be obvious therefore that there are good reasons for any book on distillation design to devote a chapter to Vapour/Liquid Equilibrium (VLE).

2.1 SATURATED VAPOUR PRESSURE

According to the molecular theory of gases and liquids, a pure liquid is in dynamic equilibrium with the vapour above it because molecules containing a certain energy level can overcome the attractive forces of a liquid and pass into the vapour. Molecules in the vapour are in constant movement and these collide with the liquid and some will stay in the liquid phase. The dynamic equilibrium so set up means that a given liquid at a given temperature will be in equilibrium with a given concentration of vapour molecules, i.e. a given vapour pressure. If an "inert gas", i.e. insoluble gas, is present in the gas phase, then, since the equilibrium set up is between the vaporised liquid molecules colliding with the surface, it is the partial pressure of the liquid components that is relevant; the pressure of an "inert gas" does not alter the situation.

The fraction of molecules in the liquid containing sufficient energy to escape is proportional to the energy level distribution. As the temperature increases this distribution shifts and the fraction possessing adequate energy increases. Since the Boltzmann energy distribution of the molecule is exponential, this means that as the temperature increases, the fraction over a specific energy increases exponentially. Hence the relationship between the temperature of a liquid and its equilibrium vapour pressure has an exponential form.

Every liquid compound has its own specific vapour pressure equilibrium curve, depending on this balance between the fraction of molecules with sufficient energy to overcome the liquid attractive forces and the number of molecules returning to the liquid from the gas. However, all components have a similar exponential form for the relationship. This relationship is basic to all distillation work, and it is recorded as experimentally determined curves, which can be transformed into parameters of a particular equation.

The equation in question is the Antoine Equation and this records the relationship between the pure saturated vapour pressure (P) and the liquid temperature
(T):

$$\ln (P) = A - \frac{B}{T + C}$$

(2.1)

This provides a surprisingly good fit to all liquids, given freedom to choose
values for the parameters A, B and C to get the best fit. The equation fit
sometimes become inadequate over wide ranges of pressure (e.g. from 0.01 bar to
1000 bar) and in this case the fit can be improved by the Extended Antoine
Equation. A frequently used extension is the 5 parameter form:

$$\ln (P) = A - \frac{B}{T + C} + DT + E \ln(T)$$

(2.2)

Alternatively the simpler form of equation (2.1) can be used by quoting
parameter sets for A, B and C, which are valid over specific pressure (or
temperature) ranges.

The simplest form of the Antoine equation is the 2 parameter form where C is
taken as a fixed value of -43.0.

2.2 LIQUID MIXTURES

Ideal Systems

Now let us assume that our liquid phase contains a mixture of two components.
We still have the same physical situation and dynamic equilibrium between the
two phases, but now the concentration of molecules in the liquid phase is different from that of the pure liquid case. For a given surface area we have only
a fraction of molecules of each component compared with the case of the pure
liquid. This fraction is the mole fraction (x) of that component in the liquid
phase. Hence a new equilibrium will be set up with a lower vapour pressure for
that component. One will expect the new vapour pressure to be reduced by that
fraction by which the liquid concentration is reduced. The experimentally
derived Raoult's law and Henry's law show this to be the case for certain liquid mixtures.

Raoult's law:

$$P_i = x_i P_i \qquad \text{for component i}$$

(2.3)

(used for high values of x - when component

i is the solvent)

Henry's law:

$$P_i = H_i x_i \qquad \text{for component i}$$

(2.4)

(used for low values of x, e.g. solubility of gases,
where H_i is the Henry's law solubility coefficient)

Hence the vapour in equilibrium with a liquid mixture will have partial vapour
pressures for each component (p_i)) defined by Equation 2.3, and the total pressure of the system will be given by Dalton's Law of partial pressures:

$$\pi = \sum_{i=1}^{n} P_i$$

(2.5)

where π is the total system pressure and n is the number of components
in the system.

Since, according to the kinetic theory of gases, the pressure is a measure of
the number of molecules in a gas phase, the vapour mole fraction (of a gas or
vapour mixture (y) is given by:

$$y_i = \frac{p_i}{\pi} = \frac{P_i x_i}{\pi} \qquad (2.6)$$

We can now see why distillation achieves separations. Since liquids have different vapour pressure temperature relationships, the vapour pressures of the
different components at a given temperature are different, and so the vapour
composition must be different from the liquid composition.

Since the differences in saturated vapour pressures are responsible for the degree of separation, a useful parameter for a system is the 'relative
volatility' (α) which is defined as the ratio of two saturated vapour pressures - usually the ratio of the higher to the lower vapour pressure.

For a binary system:

$$\alpha = \frac{P_1}{P_2} = \frac{y_1 (1 - x_1)}{x_1 (1 - y_1)} \qquad (2.7)$$

therefore

$$y_1 = \frac{\alpha x_1}{1 + x_1(\alpha - 1)} \qquad (2.8)$$

Hence the relative volatility can relate y to x. Notice however that this relationship is not very simple. The relative volatility is a useful indicator of
the difficulty of the separation. Relative volatilities less that 1.1 are very
difficult separations (e.g. separation of heavy water), those up to 1.5 are
difficult, and those over 2 are comparatively easy. The relative volatility is
used in some idealised analytical solutions to the separation problem, as will
be seen in Chapter 4.

Another convenient parameter in defining separation problems is the 'K value'.
This is simply defined as :

$$K_i = \frac{y_i}{x_i} = \frac{P_i}{\pi} \qquad (2.9)$$

The K values have to be determined by evaluating y for each particular value of
x. K is not constant but varies with temperature and concentration. It con-
tributes nothing to our understanding of the system, but it is very useful in
separating equilibrium calculations from engineering calculations, and it gives
a concise picture of the separations to be expected in a multicomponent system.

Non-ideal Systems

So far we have been discussing the 'ideal' case, with both 'ideal' liquids and
'ideal' gases. Ideality of the liquid phase assumes that when the components
are mixed together, no new interactive forces are involved - as when two types
of similar sized but differently coloured glass marbles are mixed together.
Molecules however often behave differently; different molecules may attract
each other or repel each other or react to form permanent or loosely complexed
new species. Only in specific cases do they behave as the marbles; for example
for mixtures of very similar components such as hydrocarbons, the liquid phases
behave 'ideally'.

In the general case, the equations must be written down using 'activities' (a_i) in place of the mole fractions, to maintain generality. The activity is more usually re-defined in terms of liquid activity coefficient (γ) where:

$$a_i = \gamma_i \, x_i \qquad\qquad (2.10)$$

Of course, the use of a and γ does not provide more information, they remain empirical correction factors to make the equation valid. The problem still remains; where to obtain information over non-ideality. This, basically, must be experimentally determined.

Likewise for the 'real' gaseous phase as opposed to the 'ideal gas'. To maintain validity for all gases the fugacity (f) is used in place of partial pressures, and this can again more conveniently be re-defined in terms of partial pressures and fugacity coefficients (μ), where:

$$f_i = \mu_i \, p_i \qquad\qquad (2.11)$$

again μ is an empirical correction factor to be experimentally determined.

The use of a and f, however do enable Equations (2.3),(2.5) and (2.6) to be re-written to be generally valid:

$$f_i = a_i \, P_i \qquad\qquad (2.12)$$

or

$$\mu_i P_i = \gamma_i \, x_i \, P_i \qquad\qquad (2.13)$$

$$\pi = \sum_i f_i \qquad\qquad (2.14)$$

or

$$\pi = \sum_i \mu_i \, P_i \qquad\qquad (2.15)$$

and

$$y_i = \frac{f_i}{\pi} \qquad\qquad (2.16)$$

or

$$y_i = \frac{\mu_i P_i}{\pi} = \frac{\gamma_i \, x_i \, P_i}{\pi} \qquad\qquad (2.17)$$

In this form many different thermodynamic properties can be compared. Liquid/liquid equilibrium, heats and solution, deviations from the ideal gas laws and the equation of state can be related together in terms of a and f, and to some extent experimental data from one field can be used in another. For example, heats of solution can be derived from VLE data.

2.3   SOURCES OF NON-IDEALITIES

Liquid mixtures can hardly be expected to be ideal if their chemical natures are very different and if it is to be expected that the various chemical groups will form some form of bonding or loose complexing. If the components display any heat of mixing (positive or negative), then this is a certain indication of non-ideality.

Besides these 'chemical' sources of non-ideality there are also the more physical aspect of the situation. The non-idealities of the gaseous state (molecular attraction and finite and different molecular sizes), which cause gases to be-have non-ideally also have an influence in the condensed liquid phase.

Vapour phase non-idealities come firstly from those factors that contradict the assumptions made of a perfect gas: that the molecules occupy no volume and that they do not interact with each other. These factors were noted many years ago and Van der Waals corrected the ideal gas equation for these effects.

Sources of molecular attraction in the vapour phase can be negligible for non-polar components, up to being very significant for those polar components in which attraction is so great that they dimerise in the vapour phase (e.g. acetic acid).

2.4 CORRECTIONS FOR LIQUID PHASE NON-IDEALITY

Liquid phase non-ideality - as defined by the activity coefficient - is depend-ent on concentration, being unity at mole fractions of unity and monotonically increasing or decreasing (in a binary system) as the dilution increases. Figure 2.1 shows such examples of typical γ vs x plots.

In fact this general form is defined by a form of the Gibbs-Duhem relationship which for a binary mixture is:

$$x_1 \left(\frac{\partial(\ln\gamma_1)}{\partial x_1} \right)_{T,P} + x_2 \left(\frac{\partial(\ln\gamma_2)}{\partial x_2} \right)_{T,P} = 0 \qquad (2.18)$$

This equation defines the relationship of the two curves of Figure 2.1 to each other.

The equation does not enable the value of γ to be determined, but it does en-able reported experimental data to be checked for 'thermodynamic consistency'. There clearly has to be a relationship between how 1 affects 2 and 2 affects 1 since these are 'interactions' only between 1 and 2.

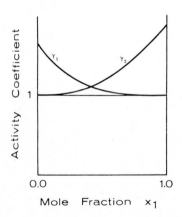

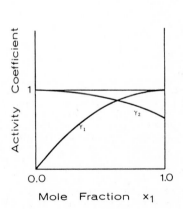

(a) Mixture with repulsive forces   (b) Mixture with attractive forces

FIGURE 2.1
TYPICAL ACTIVITY COEFFICIENT CONCENTRATION RELATIONSHIPS
FOR BINARY MIXTURES

From the definition of the Gibbs Free Energy:

$$dG = -SdT + VdP \qquad (2.19)$$

for an ideal gas at constant temperature:

$$dG = VdP = \frac{RT}{P} dP = RTd(\ln P) \qquad (2.20)$$

for a non-ideal gas component:

$$dG_i = RTd(\ln f_i) \qquad (2.21)$$

correspondingly for non-ideal liquids:

$$dG_i = RT \, x_i \, d(\ln \gamma_i) \qquad (2.22)$$

From this can be derived the relationship between the Total Gibbs Free Energy (G^E) and the liquid activity coefficients in a multicomponent system.

$$\frac{G^E}{RT} = \sum (x_i \, \ln \gamma_i) \qquad (2.23)$$

Equation (2.23) shows a relationship between $\ln(\gamma)$ and the Gibb's free energy term G^E/RT. Again this equation in itself does not define the activity coefficient but depending on the various simplifying assumptions that can be made, a number of different equations can be developed describing the form of relationship between γ and x. These different equations have all been derived by various workers and are now well known correlation equations for activity coefficients. Margules, van Laar, Wilson, NRTL and UNIQUAC are the most well known of these. These are different forms of Equation (2.23) developed using different assumptions. They all require two parameters to be fitted, except for the NRTL model which requires three for each binary system.

These equations give remarkably good fits to experimental data, once the best set of parameter values has been determined using regression analysis. Different systems have different preferred equations because of the different (and sometimes more appropriate) assumptions that each model makes.

Generalization as to the 'best' equation is not possible, except to say that the NRTL, requiring more parameters, has a better chance of producing a good fit, but it is more complicated to use and to fit.

When non-ideality becomes so great that two immiscible liquid phases develop, there is a clear distinction between the models in that only the NRTL and UNIQUAC are mathematically capable of representing a two-liquid phase region.

The choice of VLE model becomes more limited when multi-component systems, rather than simple binary systems, are to be studied. The Margules and van Laar equations are expressly for binary systems and cannot be modified for the multicomponent situation. Hence we are left with only three possibilities: Wilson, NRTL and UNIQUAC. These models have the very convenient property that the required parameters are sets of binary interactions. Thus, once the VLE data has been experimentally measured for each of the binaries in the multicomponent system, all the parameters are available to make predictions of multicomponent behaviour. Table 2.1 gives these equations in their general form.

These equations provide only the form of the correlation. The parameters generally have to be fitted to experimentally measured VLE data (y in equilibrium with x at given P and T). Much data can be found in literature, but if this is not the case for a particular system then an experimental project is necessary.

This is a time-consuming procedure requiring component purification, development of analysis, and experimentation with all the appropriate   binary pairs
over wide ranges of concentration. The value of having a predictive method for
calculating values for the model parameters without experimentation is easy to
appreciate.

TABLE 2.1

THE GENERAL FORM OF THE WILSON, NRTL AND UNIQUAC VLE EQUATIONS

Wilson $\quad \Lambda_{ij} = \dfrac{v_j}{v_i} \exp - \dfrac{\lambda_{ij} - \lambda_{ii}}{RT}$ $\qquad G^E/RT = - \sum_i x_i \ln \left(\sum_j x_j \Lambda_{ij} \right)$

$$\ln \gamma_i = - \ln(\sum_j x_j \Lambda_{ij}) + 1 - \sum_k \frac{x_k \Lambda_{ki}}{\sum_j x_j \Lambda_{kj}}$$

NRTL $\quad \tau_{ji} = \dfrac{g_{ji} - g_{ii}}{RT}$ $\qquad G^E/RT = \sum_i x_i \dfrac{\sum_j \tau_{ji} G_{ji} x_j}{\sum_\ell G_{\ell i} x_\ell}$

$G_{ji} = \exp (-\alpha_{ji} \tau_{ji})$ $\qquad \ln \gamma_i = \dfrac{\sum_j \tau_{ji} G_{ji} x_j}{\sum_\ell G_{\ell i} x_\ell} + \sum_j \dfrac{x_j G_{ij}}{\sum G_{\ell j} x_\ell} \left(\tau_{ij} - \dfrac{\sum_n x_n \tau_{nj} G_{nj}}{\sum_\ell G_{\ell j} x_\ell} \right)$

$\tau_{ii} = \tau_{jj} = 0$

$G_{ii} = G_{jj} = 1$

UNIQUAC $\quad \tau_{ji} = \exp - \dfrac{u_{ji} - u_{ii}}{RT}$ $\qquad G^E/RT = \sum_i x_i \ln \dfrac{\phi_i}{x_i} + \dfrac{z}{2} \sum_i q_i x_i \ln \dfrac{\theta_i}{\phi_i} - \sum_i q_i x_i \ln(\sum_j \theta_j \tau_{ji})$

$\tau_{ii} = \tau_{jj} = 1$ $\qquad \ln \gamma_i = \ln \gamma_i^C + \ln \gamma_i^R$

$$\ln \gamma_i^C = \ln \frac{\phi_i}{x_i} + \frac{z}{2} q_i \ln \frac{\theta_i}{\phi_i} + \ell_i - \frac{\phi_i}{x_i} \sum_j x_i \ell_j$$

$$\ln \gamma_i^R = q_i \left(1 - \ln (\sum_j \theta_j \tau_{ji}) - \sum_j \frac{\theta_j \tau_{ij}}{\sum_k \theta_k \tau_{kj}} \right)$$

$$\ell_i = \frac{z}{2} (r_i - q_i) - (r_i - 1)$$

$$z = 10$$

2.5 PREDICTIONS OF VLE PARAMETERS WITH UNIFAC

The UNIFAC method was developed in 1975 to cover precisely this need. The method enables UNIQUAC parameters to be predicted from molecular structures and is a 'group contribution' method, i.e. is the component molecules are broken down into chemical groups and the interactions between these groups is estimated and then combined to give the overall molecular behaviour.

Three specific properties must be known for each group – the molecular volume R, the molecular surface area Q and the binary group interaction parameters a_{ij} and a_{ji}.

The activity coefficient itself is the sum of the two contributions:

$$\ln \gamma_i = \ln \gamma_i^C + \ln \gamma_i^R \qquad (2.24)$$

where γ_i^C is the part due to molecular size and shape and γ_i^C is due to the molecular interactions.

The UNIFAC method contains 34 different groups, and tables exist for the constants required for these groups. The components in the system are broken down into these groups and the parameter interactions for the groups are taken to build up the total interactions between the molecules.

The R and Q data (concerned with molecular volumes and surface areas) for the different groups can be obtained from standard sources, and the interaction parameters are estimated from large quantities of existing measured VLE data. The group interaction data is therefore only available when VLE data between the various groups exists. About 70% of the interaction parameters have been evaluated. The table in the CHEMCO manual (Appendix 4) indicates which groups are available in the method and which interaction parameters exist.

When a component contains a group not amongst the 34 defined UNIFAC groups, or if any group interaction parameters are not available, then the UNIFAC method cannot be applied.

The UNIFAC method is based on the structure of the UNIQUAC VLE model, and can therefore most conveniently estimate parameter values for this model. However, UNIFAC can be used to estimate the parameters for any VLE model by firstly estimating the activity coefficients at infinite dilution for all the binary pairs in the mixture. From these infinite dilution coefficients, the parameters for any VLE model can be determined.

UNIFAC is the only predictive method for VLE that has reached general acceptance in practice. A similar method ASOC was developed independently from UNIFAC, which is again based on group contributions, but it never reached the degree of completion and reliability of UNIFAC.

When liquid non-idealities are only physical in nature, and no chemical interactions are to be expected (e.g. in the separation of hydrocarbons), then these non-idealities can be predicted by Equations of State. These equations are modifications of the ideal gas laws made to correct for non- idealities in the gas phase. These equations also predict the liquefaction point of gases and their corresponding liquid properties. Hence for non-interacting systems (e.g. non-polar), Equations of State can be used to predict liquid vapour equilibrium.

2.6 PREDICTIONS OF VAPOUR NON-IDEALITIES

For the ideal gas the following equation holds:

$$PV = nRT \qquad\qquad (2.25)$$

This equation holds adequately for most gases and vapours as long as they are not near their critical point and the pressures are not more than a few bar.

Equation (2.25) is not capable of representing the whole PV curve of a gas as shown by Figure 2.2 since this includes the change of state and information on liquid compressibility.

Modifications to the ideal gas equation enable the PV curve of Figure 2.2 to be more accurately predicted. These modified equations are called 'Equations of State', and very many have been produced and compared with experimental data. The first of these was the Van der Waals modification of the ideal gas equation by inclusion of terms representing the molar volume and molar attractive forces.

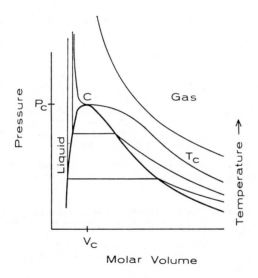

FIGURE 2.2
PVT RELATIONSHIPS FOR REAL GAS-LIQUID SYSTEMS

In general, the PVT data can be represented by the Virial Equation, which defines the compressibility factor (Z):

$$Z = \frac{PV}{RT} + \frac{B}{V} + \frac{C}{V^2} \cdots \qquad\qquad (2.26)$$

Where V is the volume of 1 mole, B is the 'Second Virial Coefficient' and C is the 'Third Virial Coefficient'

This is a polynomial to which more terms can be incorporated to extend the fit to higher pressures. For pressures less that a few bar, gases behave ideally, and the B and C constants can be zero. At higher pressures up to 15 bar, PVT data are adequately fitted if the second Virial coefficient (B) is non-zero, and for pressures up to 50 bar, the third Virial coefficient must be included to get a good fit.

The equations of state are effectively correlations for the Virial coefficients. The simplest of these is the Van der Waals equation and the simplest for use in engineering is the Redlich-Kwong equation:

$$P = \frac{RT}{V - b} - \frac{a}{T^{1/2}\,V(V + b)} \qquad (2.27)$$

Over 100 Equations of State of increasing complexity have been proposed. A more complex, well-established one is the Benedict-Webb-Rubin (BWR) equation, which requires 8 parameters:

$$P = \frac{RT}{V} + \frac{B_o\,RT - A_o - C_o/T^2}{V^2} + \frac{bRT - a}{V^3} + \frac{a\alpha}{V^6} \qquad (2.28)$$

The more constants, the wider the application of the equation. The BWR type are capable of predicting liquid conditions as well as gas and so can be used as the basis for prediction of VLE data for systems devoid of chemical interactions (i.e. non-polar systems).

For accurate work, experimental PVT data are used to evaluate all the parameters for these equations of state. However, predictions of parameter values can be made using the theory of corresponding states.

At the critical temperature and critical pressure, the PVT curve has an inflection, i.e.

$$\frac{d^2 P_c}{dV^2} = 0 \text{ and } \frac{dP_c}{dV} = 0 \qquad (2.29)$$

The Equations of State can be differentiated and relations developed for the parameters in terms of P_c and T_c. This assumes that all gases behave identically under 'reduced' conditions:-

$$T_r = \frac{T}{T_c} \text{ and } P_r = \frac{P}{P_c} \qquad (2.30)$$

For this reason, P_c and T_c are very useful data points because they can be used for predicting properties.

The theory of corresponding states is also useful for the prediction of vapour pressure. When the Antoine Equation is re-written in terms of reduced T and P:

$$\ell n(P_r) = A - \frac{B}{T_r} \qquad (2.31)$$

Then for at gases at the critical point;

$$P_r = 1 \text{ and } T_r = 1$$

hence $\qquad$ A = B

so

$$\ell n(P_r) = A(1 - \frac{1}{T_r}) \qquad (2.32)$$

It is found that different types of compound have different values for A. The 'Acentric Factor' (ω) is a definition of this A, and so, given P_c, T_c, and ω for any compound, predictions can be made of gas and liquid density, vapour pressure, fugacities, liquid activities, PVT data and VLE data for non-polar compounds.

These methods are extensively used in petroleum process design, where high pressures and non-polar compounds are involved. Critical pressure and temperature data, acentric factors and accurate PVT data exist for hydrocarbon systems.

For distillation work with these systems, Equations of State play an important role in correcting for vapour phase non-idealities at high pressure and predicting VLE data.

For 'chemical' distillation work, at pressures below 5 bar, Equations of State are of less significance. Ideality of the gas phase can be assumed, and the 'polar' nature of the compounds means that the VLE predictions by equations of state can be grossly in error.

2.7 DATA SOURCES

Before one starts to design a distillation column it is necessary to collect together the VLE data. Generally, one is looking for reported experimental data for the binary pairs of the system involved. It is most improbable that experimental multicomponent VLE data will be found for any particular system and so one generally has to be satisfied by collecting the binary pair data and predicting the effect of the multicomponent system.

There are now quite a number of collections of experimental VLE data, and so it is no longer necessary to go back to the primary literature (the research journals) to find them. The first of these collections was published in a book by Hala in 1958, and since then a progressively more modern approach has been employed. The most exhaustive of these is the Gmehling and Onken DECHEMA Series collection. This contains a number of volumes and gives all reported experimental VLE data, firstly as the raw data and then as correlations with the Margules, Van Laar, Wilson, NRTL and UNIQUAC models.

It is possible to see how well each model fits the experimental data and take the appropriately fitted parameters for the distillation design. In addition, these authors supply the same information on magnetic tape (the 'Dortmund Databank') and so access to the parameters by design programs can be made completely automatically. Figure 2.3 is a page from their printed collection.

Japanese authors have made a similar, though less exhaustive collection. Table 2.2 gives references to other useful VLE data collections.

When a search for original data fails to discover all the required experimental binary sets, the UNIFAC can be used to fill in the missing data - at least for preliminary distillation studies. Published tables of the required UNIFAC Q, R and interaction parameters are available (Fredenslund, 1976) and reports in the research journals contain the updates to this material.

If the UNIFAC parameters do not exist for the published groups involved in the system, or if preliminary studies with UNIFAC show that a system presents a serious separation problem, then it will be necessary to resort to experimental measurements of the VLE.

Laboratory measurement of VLE must be carried out in a so-called equilibrium still. This is a device for boiling a liquid mixture, containing the vapour and recycling it to the boiling liquid. When equilibrium is attained, samples are

(1) METHANOL   CH4O
(2) WATER   H2O

+++++ ANTOINE CONSTANTS   REGION +++++
(1) 8.08097 1582.271 239.726   15- 84 C   METHOD 1   CONSISTENCY +
(2) 8.07131 1730.630 233.426   1- 100 C   METHOD 2   +

PRESSURE= 760.00 MM HG (1.013 BAR)

LIT: GCON J.,TABOADA C.
     AN.REAL.SOC.ESPAN.DE FIS.Y QUIM.,55B(3),255(1959).

CONSTANTS:	A12	A21	ALPHA12
MARGULES	0.7279	0.6174	
VAN LAAR	0.7287	0.6237	
WILSON	-105.9234	648.0054	
NRTL	-11.5842	545.3615	0.3011
UNIQUAC	199.3374	-109.1007	

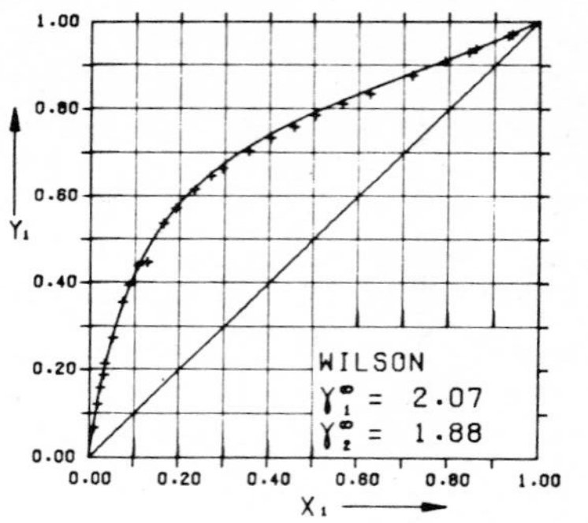

EXPERIMENTAL DATA			MARGULES		VAN LAAR		WILSON		NRTL		UNIQUAC	
T DEG C	X1	Y1	DIFF T	DIFF Y1	DIFF T	DIFF Y1	DIFF T	DIFF Y1	DIFF T	DIFF Y1	DIFF T	DIFF Y1
97.44	0.0115	0.0678	-0.66	-0.0089	-0.66	-0.0090	-0.66	-0.0090	-0.67	-0.0085	-0.64	-0.0097
96.92	0.0198	0.1214	0.08	-0.0038	0.08	-0.0038	0.08	-0.0038	0.06	-0.0033	0.11	-0.0048
95.82	0.0258	0.1589	-0.17	0.0017	-0.17	-0.0017	-0.17	-0.0017	-0.19	0.0023	-0.13	0.0006
95.06	0.0330	0.1882	0.04	-0.0043	-0.04	-0.0043	-0.05	-0.0043	-0.03	-0.0038	0.08	-0.0055
94.13	0.0357	0.2145	-0.54	0.0095	-0.54	-0.0095	-0.53	-0.0096	-0.55	0.0100	-0.49	-0.0083
92.24	0.0525	0.2746	-0.40	-0.0002	-0.40	-0.0001	-0.40	0.0002	-0.41	0.0002	-0.36	-0.0011
90.00	0.0740	0.3560	-0.38	0.0083	-0.39	0.0085	-0.39	0.0091	-0.38	0.0085	-0.35	0.0079
88.57	0.0872	0.3950	-0.58	0.0095	-0.59	0.0098	-0.60	0.0107	-0.57	0.0097	-0.56	0.0094
86.93	0.1079	0.4400	-0.49	0.0036	-0.49	0.0039	-0.52	0.0052	-0.47	0.0038	-0.48	0.0040
85.37	0.1289	0.4476	-0.49	-0.0323	-0.50	-0.0320	-0.54	-0.0303	-0.47	-0.0320	-0.51	-0.0314
83.38	0.1635	0.5370	-0.28	-0.0013	-0.29	-0.0010	-0.35	0.0012	-0.26	-0.0008	-0.32	-0.0002
81.95	0.1912	0.5724	-0.20	-0.0040	-0.22	-0.0036	-0.28	-0.0011	-0.19	-0.0031	-0.27	-0.0021
80.25	0.2327	0.6162	0.03	-0.0069	0.01	-0.0066	-0.07	-0.0037	0.03	-0.0055	-0.06	-0.0048
79.06	0.2684	0.6483	0.25	-0.0079	0.24	-0.0076	0.14	-0.0045	0.23	-0.0061	0.16	-0.0057
78.14	0.2942	0.6658	0.25	-0.0112	0.23	-0.0110	0.14	-0.0078	0.23	-0.0092	0.15	-0.0091
76.52	0.3524	0.7044	0.42	-0.0128	0.41	-0.0127	0.32	-0.0096	0.39	-0.0105	0.32	-0.0109
75.34	0.4021	0.7341	0.56	-0.0120	0.55	-0.0121	0.46	-0.0092	0.52	-0.0096	0.46	-0.0106
74.22	0.4543	0.7595	0.66	-0.0135	0.66	-0.0136	0.58	-0.0110	0.61	-0.0110	0.58	-0.0124
73.21	0.5022	0.7853	0.66	-0.0100	0.67	-0.0103	0.60	-0.0080	0.62	-0.0077	0.60	-0.0094
71.95	0.5628	0.8123	0.57	-0.0092	0.58	-0.0095	0.54	-0.0077	0.53	-0.0072	0.53	-0.0091
70.90	0.6243	0.8350	0.62	-0.0117	0.63	-0.0120	0.60	-0.0105	0.59	-0.0100	0.59	-0.0119
69.15	0.7173	0.8773	0.41	-0.0062	0.42	-0.0064	0.41	-0.0054	0.39	-0.0050	0.40	-0.0066
68.07	0.7898	0.9098	0.46	-0.0023	0.47	-0.0023	0.47	-0.0017	0.44	-0.0015	0.46	-0.0026
67.17	0.8426	0.9300	0.35	-0.0032	0.36	-0.0033	0.37	-0.0027	0.35	-0.0028	0.35	-0.0035
66.90	0.8574	0.9385	0.30	-0.0007	0.31	-0.0007	0.32	-0.0003	0.30	-0.0004	0.30	-0.0010
65.73	0.9295	0.9682	0.18	-0.0011	0.19	-0.0010	0.19	-0.0007	0.18	-0.0009	0.18	-0.0011
65.71	0.9380	0.9712	0.28	-0.0017	0.29	-0.0016	0.29	-0.0014	0.28	-0.0016	0.29	-0.0017
MEAN DEVIATION:			0.38	0.0073	0.39	0.0073	0.37	0.0063	0.37	0.0065	0.36	0.0069
MAX. DEVIATION:			0.66	0.0323	0.67	0.0320	0.66	0.0303	0.67	0.0320	0.64	0.0314

WILSON
Y_1^∞ = 2.07
Y_2^∞ = 1.88

FIGURE 2.3
EXAMPLE OF VLE DATA REPORTED IN THE DECHEMA DATA SERIES

taken of the boiling liquid and condensed liquid, and these are analysed to provide one VLE x/y point. Temperatures and pressures are measured, thereby completing the data set for that point.

TABLE 2.2
SOME REFERENCES TO VLE LITERATURE DATA COLLECTIONS

---

(1) VAPOUR PRESSURES ONLY

    Boublik T., Fried V., Hala E., 1984. Vapour Pressure of Pure Substances, Elsevier.

    Gallant R.W., 1969. Physical Properties of Hydrocarbons, Gulf Publishing.

    Perry R.H., Chiltern C.H.,1973. Chemical Engineers Handbook, McGraw-Hill.

    Timmermanns J.,1950. Physico-Chemical Constants of Pure Compounds, Elsevier

    Yaws C.L.,1977. Physical Properties of Industrially Important Chemical Compounds, McGraw-Hill.

(2) VAPOUR/LIQUID EQUILIBRIA

    Gmehling J.,Onken U.,1977. Vapour-liquid Equilibrium Data Collection, DECHEMA Chemistry Data Series, DECHEMA.

    Hala E.,Pick J.,Fried V.,Vilin O.,1958. Vapour-Liquid Equilibrium, Pergamon.

    Hirata, M., Ohe S., Nagahama K.,1975. Computer Aided Data Book of Vapour-Liquid Equilibria, Elsevier.

This simple explanation hides the pitfalls and difficulties that are present in obtaining accurate data. Firstly, the components must be very pure; otherwise impurities will also separate in the still and provide false data. Secondly, there must be no partial fractionation or wetted-wall effect between the boiler and condenser that could provide additional enrichment. Much thought and development have gone into the design of equilibrium stills to ensure that exactly one plate separation is achieved. The apparatus must also be capable of operating at high pressures and also under vacuum. Hence, well-engineered stainless steel equipment is preferred to a glass still.

It is not the aim of this chapter to give guidance on the experimental determination of VLE data, but only to point out that this work is not easy: it takes careful preparation and only two or three points per day can be evaluated. If a system has to be measured, one can reckon on a month of work unless a very well equipped specialised laboratory is available.

Once the x/y and vapour pressure data have been determined experimentally,it is necessary to fit the appropriate VLE model to enable the results to be summarised suitably for calculation work.

The Antoine Equation has to be fitted to the vapour pressure/temperature data. This is comparatively simple, and it is possible to convert the problem into a linear problem, which further simplifies the fitting.

The VLE model fitting is a more complicated problem. Accurate data is required for design, and a number of VLE models are available. The best fitting model is to be preferred, thus all the models should be tried to enable the choice to be made. Since the models are all good representations of the data, there is the opportunity to use the more sophisticated regression techniques to get a greater confidence level for the fitted parameters (e.g. maximum likelihood as opposed to simple sum of squares fitting).

The problem is multi-response (measured are x, y, T and P) and T and y, or P and y can be fitted. Furthermore, the range of y is large (e.g. 0.00001 to 0.99999), and straightforward fitting yields poor results at low concentrations. A modified response to give appropriate weightings for future use in design is required.

These problems are generally solved in computer packages specifically developed for fitting VLE data. Fitting is usually done to measured binary data, and the ternary or higher numbers of components predicted using combinations of the binary pair parameters with the VLE model. A check on the accuracy of such predictions is then made by measuring a few multicomponent data points and comparing the measurements with the predictions to see whether this approach is adequate.

A set of such programs called VLESET is available from EURECHA as part of the same distribution scheme as CHEMCO and DISTILSET, which are detailed in the appendices.

Alternatively to measuring VLE data specifically in a laboratory, they may be determined by parameter fitting results of a pilot distillation, should a pilot distillation be necessary for other reasons. This procedure is viewed with horror by traditional VLE men, but, as reported in Chapter 9, surprisingly good data can be obtained in this way.

2.8 DATA BANKS

The design of distillation columns requires a great deal of physical property data: vapour pressure data and VLE data for the separation itself, liquid and vapour density and viscosity data, and surface tension data to design the hydrodynamics of the column for plate or packing selection, and finally liquid and vapour enthalpy data for the heat loads on the column.

Since distillation design is carried out by computer, it is clearly both a laborious and error-prone procedure to enter the coefficients for all this data by hand for every distillation computer run. It is much more sensible to have all this physical property data previously stored in the computer in a 'physical property data bank', which can be called by the distillation program and the required component data transferred automatically. During the 1970's a number of these data banks were developed. Some were mounted on computer networks so that the data could be accessed by terminal from any telephone,and others were mounted on the same computer as the engineering programs to enable direct access to be easily made. Some of these banks were privately developed to support in-house engineering programs, and some ventures were intended as publicly available banks which could be used for a fee or purchased outright.

Table 2.3 lists some of these data banks under development in 1978.

2.8a DATA BANK STRUCTURE

Different data banks have different structures, which are primarily dictated by whether the data set is stored in a direct access or sequential file. For banks with direct access files, systems software locates directly the relevant record and loads the data into the working space. For systems without the direct access facility, all files have to be read sequentially and the irrelevant records skipped until the appropriate record is found. For sequential files, where the single access takes longer, extra management is required by the data

TABLE 2.3
SOME MAJOR COMPUTER-BASED PHYSICAL PROPERTY DATA BANKS

Bank Name	Organisation
SDC	DECHEMA
TISDATA	DSM
DATABANK	ICI
PPDS	Inst. of Chem. Eng.
FLOWTRAN	Monsanto
EPIC	Liege University
CHEMCO	EURECHA
PHYSCO	Milan Polytechnic
UHDE SDC	Uhde
CHESS	Washington University
ASPEN-PLUS	AspenTech

bank software to ensure that the sequential file is read the minimum number of
times.

For such banks, a pre-calculation phase is required,which informs the bank what
data is going to be called for. This information is stored, but no retrieval is
done until the first read call for data. At this point the file is read through
once and all the data notified as going-to-be-read is retrieved from the file.
With this facility programmed into the data bank software, the sequential access time becomes smaller than the total time for direct access. For the
engineer, direct or sequential access therefore are equally suitable.

With both types of access, it is important that once the data has been
retrieved it remains available without further need for re-access.   Therefore
this data is stored in the data bank program itself and thus it is easy to use
that same data as often as required without re-accessing the data file. This
feature is important for data banks that work in conjunction with engineering
programs, because an engineering program makes many thousands of data calls during a single execution.

To   the   user, the data bank requires only a specification of the components and the property required (at specified conditions) to be made. The data
bank software - which is by no means trivial - manages the call, collects the
appropriate parameters, either from the file or from the local store, computes
the appropriate property value and returns the solution.

Some data banks also have a predictive part where they predict component
properties, not from parameters from measured data, but by predictive equations
based on equations of state or group contribution methods. Such methods are
widespread for the petroleum and petrochemical systems but are less reliable
for chemical, polar systems.

2.8b FLASH CALCULATIONS WITH DATA-BASES

It is usual for data banks also to have the necessary software to predict VLE
data. That is, from their stored data they return the results of a 'flash' calculation giving equilibrium liquid and vapour compositions using any specified VLE model.

Flash calculations can take many forms. In the general 'flash' situation, where
we are feeding a mixture at one temperature and pressure and taking away two

phases at a different temperature and pressure we are handling a system with the following variables:

$$\text{feed:} \quad z_i, \ T_F, \ P_F, \ Q$$

$$\text{output:} \ x_i, \ y_i, \ T, \ P, \ V$$

where Q refers to the heat input to the system from an external source and V is the vapour fraction leaving the flash, and z_i is the inlet mole fraction.

Hence, depending on what variables are to be set as dependent and independent, we get different types of flash calculation. The six most familiar are given in Table 2.4.

TABLE 2.4
DIFFERENT TYPES OF FLASH CALCULATIONS

Description	Defined Conditions	Calculated Conditions
1 Vapour pressure and composition over a given liquid	x,T	P,y
2 Boiling point of a liquid	x,P	T,y
3 Dew point of a vapour	y,P	T,x
4 Partial vaporisation	z,P,V	x,y,T,Q
5 Isothermal flash	z,T,P	x,y,V,Q
6 General flash (also adiabatic flash when Q=0)	z,P,Q	x,y,T,V

Symbols:
   z is feed composition
   V is vapour fraction
   Q is enthalpy change by flash

The calculation of these six flash types for multicomponent systems generally cannot be done analytically; the solution is found by iterative methods. Since flash calculations are central to much engineering work (particularly, boiling point and bubble point calculations are very important in distillation work), much effort has been spent in deriving methods that are quick and reliable.

The basic equations involved are:

The relation between liquid and vapour composition:

$$y_i = \frac{\gamma_i \ x_i \ P_i}{\pi} \tag{2.33}$$

The relation between liquid activity coefficient and temperature and composition:

$$\gamma_i = f(T,x) \tag{2.34}$$

The relation between saturated vapour pressure and temperature:

$$\ln(P_i) = A - \frac{B}{C + T} \tag{2.35}$$

and the condition that mole fractions must add up to unity:

$$\sum y_i = 1.0 \text{ and } \sum x_i = 1.0 \tag{2.36}$$

Only in the case of type 1 flash calculation of Table 2.4 (the vapour pressure over a given liquid composition at a given temperature) can the equations be solved directly.

In this case, Equation (2.34) gives the activity coefficient directly, Equation(2.35) gives the saturated vapour pressure of each component and Equation (2.33), in conjunction with Equation (2.36) gives the total vapour pressure of the mixture (π) and vapour composition.

For the boiling point calculation, the temperature must be found where the total vapour pressure equals the system pressure. Hence T is not known and γ and P cannot be calculated without iteration.

In general, an initial estimate of this temperature is made, γ , P,y and π calculated, and the residual, expressed as:

$$R = \sum y_i - 1 \tag{2.37}$$

is reduced to zero (or within an acceptable tolerance of zero, defined by the accuracy required by the calculation) by repeating estimates of T, using the secant method to find the root.

This calculation is central to all distillation calculations, it must be done with minimum iterations. Hence the function should be as linear as possible. By making the transforms:

$$\text{P is expressed as } \ln(P)$$
$$\text{and} \quad \text{T is expressed as } 1/T$$

The resulting function is very linear, and this results in much quicker convergence than when using the secant with the untransformed variables. Convergence is usually achieved in 3 iterations.

The calculation of flash types 3 to 6 are even less direct, because the liquid composition is not given. This results in n-1 unknowns in addition to the unknowns that the flash is solving for.

This liquid composition is determined by the solution of Equations (2.33) to (2.36) whilst maintaining a material balance over the system:

$$z_i = y_i V + x_i(1 - V) \tag{2.38}$$

assuming a feed of 1 mol, of mole fraction z, with a resulting vapour fraction V.

It is generally found to be more efficient and reliable to solve these equations as a nested set of iterations rather than let a multi-dimensional mathematical root finder (e.g. Newton-Raphson or multi-dimensional secant) locate the solution.

Firstly, for a given T and V, a liquid composition x* is found which satisfies the relationship between liquid and vapour composition:

$$K_i = \frac{y_i}{x_i} = \gamma_i\, P_i \qquad (2.39)$$

This requires an iterative procedure because x must be known before γ can be calculated. This is usually done by a form of repeated substitution, with an additional normalisation step, since this is necessary for some of the VLE models, i.e. forms of Equation (2.34), that are used:

$$x_i = \frac{x_i^*}{\sum x_i^*} \qquad (2.40)$$

This iteration step converges on a constant value of K. However, since T, V or P are not yet located, generally:

$$x^* = 1 \quad \text{and} \quad y = 1$$

the outer convergence loop now identifies the values of T,P or V which satisfies the conditions defined by Equation (2.36). Flash type 4 of Table 2.4 searches for T, and type 5 for V. This single variable search is generally done using the secant method, but again a more linear form of the equations can be used leading to a quicker, more robust solution by transformation. The combination of the two conditions given by Equation (2.36) can be achieved by using the expression:

$$\frac{\sum x_i}{\sum y_i} = 1.0 \qquad (2.41)$$

The evaluation of specific examples has shown that transformations giving near-linear characteristics are obtained when the solution to

$$\ln\left(\frac{\sum x_i}{\sum y_i}\right) = 0$$

is sought by searching with 1/T or ln(P). (See Vidhog B.G. Chemdata Proceedings, Helsinki 1977, Science Press, Princetown, N.J.)

Forthe Flash type 6, the conditions achieved when a defined energy change (Q), is requested. The adiabatic flash is a special case of this type where Q = 0. This type requires a third convergence loop to achieve the energy balance.

$$\sum y_i\, V\, H_i^V + \sum x_i (1 - V)\, H_i^L = \sum z_i\, H_i^F + Q \qquad (2.42)$$

where   Q is the energy addition to the system per mol of
        feed (kJ/kmol)
and   H are the appropriate liquid or vapour component
        molar enthalpies.

Three distinct regions for the solution exist:

- Subcooled liquid: where V = 0 , and T is found to satisfy Equation (2.42)

- Superheated vapour: where V = 1 , and T is found to satisfy Equation (2.42)

- Two phase system: where 0 < V < 1.0 and V is found, via Flash type 4, to satisfy Equation (2.42).Because of azeotrope formation and the case of a single-component system, it is not possible to search for T as in the above two cases.

These searches are done using the secant method without further transformations, since the system is fairly linear.

Clearly the Flash Package is not a trivial part of the data bank software.

2.8c TEMPORARY ADDITION OF MISSING DATA

The situation often arises when the data file does not contain information on
all components in a mixture under study. But since the data bank is prepared to
calculate mixture properties and also VLE data, it is necessary to make provisions for these facilities to be used even when the complete data set is not on
a file.

This is achieved by supplying facilities for adding data temporarily, simply
for the duration of the computer run. On termination of the run the added
data are lost.

It is necessary to consider these data to be temporary because they may not
have passed the accuracy tests laid down for the permanent incorporation of
data into the data file. In fact, the data used in lieu of missing data may be
purely imaginary. For example a component may be missing, it may be known to
have a low boiling point, and its composition is low so it will behave as a
light and have no effect on heat balances and flow volumes. Under these particular circumstances an adequate set of parameters for the vapour pressure,
enthalpy and density can be created without reference to any measured data,
which enables the bank to calculate adequately the total system - but on no account should such data be stored permanently!

2.8d ACCESS TO DATA BANKS

The 'data bank' so far discussed is a set of subroutines and a data file which
can be called to deliver any required data point. So there needs to be a calling program available to use the data bank. Two possibilities exist: either the
calling program is the engineering program needing the data, in which case a
standard call will provide the main program with the required data and the engineering calculation can proceed; or a main program must be provided to return
and display the results so that the engineer can be presented with a display of
the data that he requires.

In the first instance there is a very minimum of interaction between the engineering program and the data bank subroutine. In the CHEMCO data bank, whose
manual is given in Appendix 4, the only statement required to provide, for
example, the enthalpy of a mixture is:

           CALL CHEMCO (2,57,NORD,NCOMP,X,Y,T,P,ENT,IER)

which delivers to the calling program the enthalpy of the mixture of NCOMP components whose component numbers are defined by the array NORD, of liquid composition x at T(K) and P(bar) in the variable ENT. (The data bank code for vapour
enthalpy of a mixture is 57).

Because of this simplicity in calling the data, there is no great problem in
converting engineering programs to call on a data bank for its data, or in
changing one data bank for another, but keeping the same engineering program.

With regard to the second accessing possibility of a data bank - to use a main
program to print the data on a terminal or in printed form - again various possibilities exist. The program can deliver simply a point value, or it could
present a complete table (vs temperature for example), by its own repeated call
from the data bank. Ideally, both possibilities should exist, as it is annoying
to wait (and pay for) the whole table to be produced when only a single point

is needed, but it is also annoying to make multiple runs when you need an in-
house document defining the physical properties of a system over a range of
temperatures.

There are good reasons for having such main programs interactive, since different properties require different information, and mistakes in the sequence of
the input data are frequent unless a prompting system is available.

2.9 PSEUDO-COMPONENTS

The distillation of natural products (for example crude oil or liquified coal)
requires a technique for handling hundreds of components in the system.
Distillation design methods analyse a limited number of defined components; natural materials contain a range of different, unidentified components. A crude
oil sample or any oil fraction is characterized by a standard ASTM boiling
point curve. The sample is distilled in a defined apparatus at a defined heating rate, and the volume fraction recovered versus temperature recorded to give
the ASTM boiling point curve for that sample (see Figure 2.4).

This curve is made up of very many different components of different boiling
points, compounded with the low degree of fractionation of the ASTM test. A
high degree of fractionation would give a curve composed of very many steps for
a mixture of a few components, or a modified curve when very many components
are present (see true boiling point curve (TBPC) on Figure 2.4). Because it is
difficult in practice to obtain a TBPC, the ASTM standard test procedure has
been developed; this is easier to perform and can still be used   to
'characterize' our sample to enable design calculations to be performed.

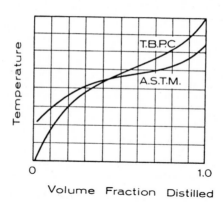

FIGURE 2.4
TYPICAL ASTM BOILING POINT CURVE

2.9a CHOOSING THE NUMBER OF COMPONENTS

The curve is segmented into a number of different stages, and each segment is
then assumed to be a pure component whose boiling point is the mid-tempera-
ture of that segment. How many segments are chosen depends upon the specification of the final separated products of the system.

If we are considering a single distillation, we have to ask ourselves how
accurate do we require the specification. If the quantity and properties of the
low boiling impurities must be known in the distillation, the accuracy
with which these are known must be transmitted by having this accuracy in the

original feed. For example, if we need to known the % boiling below 100°C, and % below 130°C in a 170°C fraction - even if these % are less than 1% - then we must define appropriate fractions ('pseudo-components') with these boiling ranges in the feed.

There are no rules for selecting the numbers of pseudo-components to choose to represent the system - it depends entirely on the design problem. For simple single distillation purposes one may need from 5 to 10 components, for simulation of complete processes, in which the feed eventually may produce 6-8 final product streams as in oil refining, a much finer description is necessary and 25 to 50 pseudo-components may be required.

2.9b DEFINING THE PSEUDO-COMPONENT PROPERTIES

Having defined the boiling point for each pseudo-component, the next step is to define this component in sufficient detail for it to be used in the normal design procedures so that, finally, the plant dimension can be given for the required separation.

We would like to have a vapour pressure curve, VLE data (or deviations from ideality), density, molecular weight, vapour and liquid viscosities and thermal conductivities.

If we assume that all the pseudo-components belong to the same chemical species - all aliphatic hydrocarbons, for instance - then we can interpolate between the homologous series and predict what the equivalent physical property data would be (notice that we are really dealing with pseudo-components because the actual components may not exist, but interpolated properties will suffice).

Unfortunately, in petroleum fractions all components are not the same chemical type, and crude oils from different sources have different chemical constituents; the main difference being the proportion of aromatic, rather than aliphatic hydrocarbons present.

Successful pseudo-component property prediction methods are therefore based upon defining the pseudo-component in terms of its boiling point, and a second property which defines its 'aromaticity'. One such property is its C:H ratio, which is higher for aromatic compounds.

Nomographs have been developed to use these two data points to define pseudo-components accurately enough for distillation design work. Figure 2.5 shows a typical way in which such data can be presented.

From a straight line on Figure 2.5 joining any two known properties, all the remaining properties can be read off. Instead of the boiling point plus C:H ratio, we could have boiling point plus molecular weight, or boiling point plus density, and our pseudo-component can be characterized.

As interest grows in liquefaction of coal, we see liquids being produced with rather different chemical properties, and this reflects itself in the definition of pseudo-components. The relationships given by Figure 2.5, which are based on petroleum fractions must be modified to be applicable to coal liquid fractions because the chemical nature of the components is different.

Those methods of predicting physical property data (usually hydrocarbon data) from critical properties, acentric factors, etc. are also applicable for hydrocarbon pseudo-components. Hence, from the pseudo-component characteristics the appropriate critical properties and acentric factors can be defined, and so the pseudo-components can be used with design methods connected to predictive data banks in a perfectly normal way.

Pseudo-components can therefore be used with data banks by inputting the characteristic properties of the components. This procedure, however, does not make use of the main advantage of the data bank,which is the storage of data so that components can be called up without the inputting of any data other than their name.

This facility can be provided also for pseudo-component by enabling the component to be identified to the bank by a code name containing its boiling point and density. The data bank can then do the remaining calculations and provide any data on request.

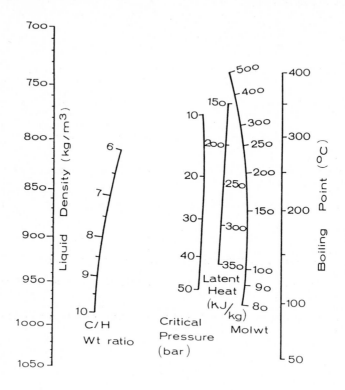

FIGURE 2.5

NOMOGRAPH FOR THE SELECTION OF PSEUDO-COMPONENT PROPERTIES

Naturally, during the years of petroleum refining, the subject has become very developed and within a company all crudes are already identified by appropriate sets of pseudo-components, and the engineer can design with them in exactly the same way as if they were mixtures of real components.

2.10 FURTHER READING

Fredenslund A., Gmehling J., and Rasmussen P., 1977. Vapour-Liquid Equilibria using UNIFAC, Elsevier.

Fredenslund,A., Rasmussen P. and Mollerup J., Thermophysical and Transport Properties for Chemical Process Design. Eng. Foundation Conference, July 1980 – very competent survey of estimation methods for thermodynamic properties for engineering design.

Holland C. D. 1981. Fundamentals of Multicomponent Distillation, McGraw-Hill
- an excellent chapter on prediction of physical properties of non-polar
  compounds with Equations of State.

Rasmussen, P., 1980. Data Banks for Chemical Engineers, Lyngby Tech. Univ.
- a good introduction to the subject.

Smith J.M and Van Ness H.C., 1975. Introduction to Chemical Engineering
Thermodynamics, McGraw-Hill
- a very readable treatment of the fundamentals.

References to VLE data are summarised in Table 2.2

2.11 QUESTIONS TO CHAPTER 2

1) Why do vapour compositions generally differ from the liquid compositions
   they are in equilibrium with?

2) What parameters can be used to give a qualitative and convenient measure of
   these differences in liquid and vapour composition?

3) What is the liquid activity coefficient, and what models exist to describe
   it?

4) How is UNIFAC different from the Wilson and NRTL models?

5) At what pressures do vapour non-idealities become important, and how can
   they be handled?

6) What is a physical property data bank, and how are they used?

7) What are the different forms of flash calculation, and why are transformations necessary for their solution?

8) Why are boiling point curves and pseudo-components needed for petroleum
   refinery design?

Chapter 3

BINARY DISTILLATION

3.1 THE CALCULATION OF THE NUMBER OF STAGES

3.1a GRAPHICAL METHODS

The classical way of teaching binary distillation is by the McCabe-Thiele diagram. An x/y equilibrium plot of the system in question is constructed, and the top and bottom operating lines and the 'q' line representing the heat content of the feed is super-imposed on the x/y plot. The number of theoretical stages above and below the feed can then be stepped up between the operating line and the equilibrium line (as shown in Figure 3.1). Any standard text can be consulted for details of its construction and proof.

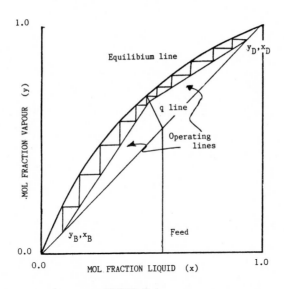

FIGURE 3.1
MCCABE-THIELE DESIGN METHOD

The method is certainly very good for teaching. It gives a visual understanding of easy and difficult separations and shows particularly well the minimum reflux situation, where an infinite number of plates is needed to achieve the required separation (see Figure 3.2).

It also shows how some equilibrium curves can cause minimum reflux conditions, where the 'pinch' (i.e. where many plates cause virtually no concentration change) occurs at other than the feed point (Figure 3.3).

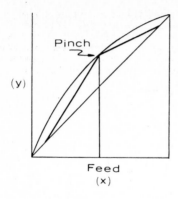

FIGURE 3.2
MINIMUM REFLUX WITH   PINCH
AT FEED POINT

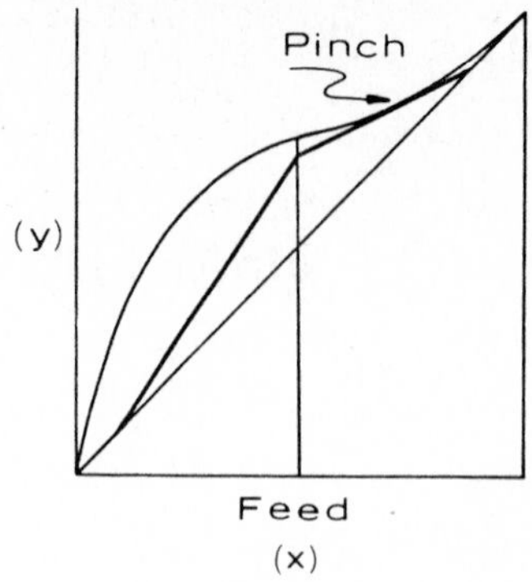

FIGURE 3.3
MINIMUM   REFLUX   WITH PINCH
AT OTHER THAN FEED POINT

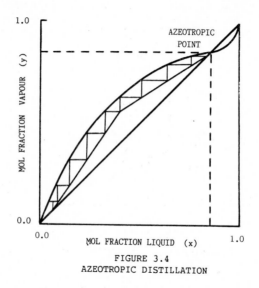

FIGURE 3.4
AZEOTROPIC DISTILLATION

It is also easy to see the problem of azeotropes and the fact that no
distillation condition can take compositions beyond the azeotropic point, where
the equilibrium line crosses the diagonal (Figure 3.4).

The McCabe-Thiele construction is derived from from an algebraic analysis of
the distillation process, and it is therefore equally possible to determine
the number of plates required by an algebraic calculation followed by a numerical analysis.

3.1b NUMERICAL METHODS

For the sake of clarity and simplicity in this analysis let us make the
'constant molal overflow' assumption. This means that that the vapour molar
flow on each plate is the same and likewise the liquid molar flow is constant

- except for feed and sidestream changes. Constant molal overflow assumes that there is no sensible heat flow in the column(i.e. temperature differences involve negligible sensible heat flows), all components have the same latent heat, and there is no heat loss from the column. Usually these are acceptable assumptions when one is determining only the number of plates for a separation. Under these circumstances, the flows around the column can be calculated by mass balance.

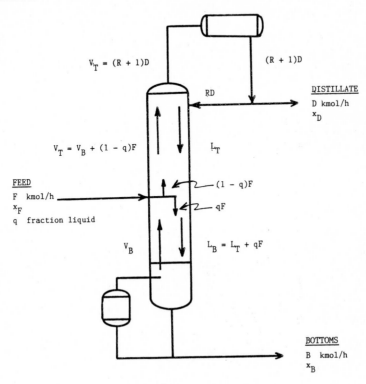

FIGURE 3.5
FLOWS AROUND A DISTILLATION COLUMN

Figure 3.5 indicates the relationship between the various total flows (assuming constant molal overflow). For a binary distillation with a specification setting the composition of top and bottom products, it is possible to carry out component mass balances over the whole column and define flows D and B in terms of x^D, x^B, and x^F.

For a component 1 balance

$$Fx_1^F = Dx_1^D + Bx_1^B \tag{3.1}$$

and a component 2 balance:

$$Fx_2^F = Dx_2^D + Bx_2^B \tag{3.2}$$

eliminating B from these two equations we can define D in terms of known (specified) variables.

$$\frac{Fx_1^F - Dx_1^D}{Fx_2^F - Dx_2^D} = \frac{x_1^B}{x_2^B} \tag{3.3}$$

then

$$Fx_1^F x_2^B - Dx_1^D x_2^B = Fx_2^F x_1^B - Dx_2^D x_1^B \tag{3.4}$$

Hence

$$D = F \frac{x_1^B x_2^F - x_1^F x_2^B}{x_1^B x_2^D - x_1^D x_2^B} \tag{3.5}$$

Once D has been numerically evaluated, B can be determined from equation (3.1).

$$B = \frac{Fx_1^F - Dx_1^D}{x_1^B} \tag{3.6}$$

Variables B, D, F, and q are all known from the definition of the problem, and hence all the flows around the column, including the internal column flows V and L below and above the feed are known.

If we now consider the first plate and boiler of a distillation column such as shown in Figure 3.6, then we can write down relationships for the liquid and vapour compositions on the plates.

Using the convention that the stream leaving plate n is subscripted n (boiler being plate 0), then we have from each plate x_n, y_n, L_n, V_n, where x and y are liquid and vapour mole fractions, and L and V are molar flow rates.

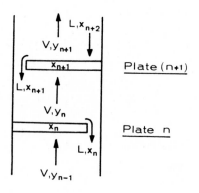

FIGURE 3.6
FLOWS AROUND A TWO PLATE SECTION OF A COLUMN

Starting with plate 1, we can write down a mass balance for component 1:

Total kmol/s in = Total kmol/s out

hence

$$Lx_2 + Vy_0 = Lx_1 + Vy_1 \qquad (3.7)$$

(where the subscripts refer to the plate number, not the component)

y_1 is in equilibrium with x_1 because in our theoretical analysis we are assuming that the plate behaves ideally with a plate efficiency of unity. Hence y_1 can be calculated from x_1 using Raoult's law for the ideal case or, for the non-ideal case

$$y_1 = \frac{\gamma_1 \, x_1 \, P_1}{\pi} \qquad (3.8)$$

Since y_0 and x_1 are known from the boiler composition; assuming a recirculation-type boiler then both are equal to the bottom product specification; and since L and V are known from Figures (3.5) and (3.6), the only unknown in the Equation is x_2, which we can find by rearranging Equation (3.7).

$$x_2 = x_1 + \frac{V}{L} \, (y_1 - y_0) \qquad (3.9)$$

This procedure can now be repeated for the following plates until the feed composition is reached. This then represents the ideal position for the feed point, and so appropriate values for L and V above the feed can be used in the Equation equivalent to (3.9) and the plate calculation repeated until the required distillate composition is reached.

In this way the number of plates can be determined by a plate-to-plate calculation involving no iteration. This is possible only for binary distillation because the top and bottom compositions for all components can be defined in advance of the calculation. We will see in later chapters that this is not possible for multicomponent systems and then iterative procedures must be used.

The basic Equations (3.8) and (3.9) can be easily programmed to form a binary distillation design program, and we have our first example of CAD in distillation.

It should be emphasised that the simplest form of numerical binary distillation is indeed very simple. The program requires only 16 program statements and this can be condensed into 6 lines of FORTRAN coding.

However since we have such a clear method as the graphical McCabe-Thiele available for binary systems, is there any point in having a computerised binary design method, even if it is simple?

There are a number of deficiencies with McCabe-Thiele that can only be rectified by a numerical solution, as the following paragraphs will show.

VLE AVAILABILITY

McCabe-Thiele looks simple only because the VLE x/y curve can be supplied. If it were necessary to calculate this curve from VLE models and Antoine parameters, the calculation effort would be similar to that required to calculate the distillation numerically. It would have lost its simplicity.

Reported measurements of VLE extend over varying pressures, temperatures, compositions: some isothermal, some isobaric. A VLE model can correlate these different data and present a best fit to it by a set of parameters. To rely upon one measured equilibrium curve does not make use of the other measured data. Furthermore, it is pure chance if the data has been measured under conditions of pressure and composition required by the distillation problem. In particular, effects of pressure on separation, or corrections due to pressure drop, cannot be investigated with a McCabe-Thiele diagram with its single equilibrium curve.

PRESENTATION OF PROFILES

McCabe-Thiele gives a visual interpretation of steps wide apart or close together. This infers that concentration changes per plate are correspondingly large or small. This information is given in tabular form by a plate-to-plate analysis, a slightly different but professionally more acceptable way of presenting the information. More important is the calculation and presentation of the temperature profile by the plate-to-plate method. This information is not available with the McCabe-Thiele method. It will be shown later in this chapter that understanding the profiles is essential to the development of a good distillation design.

McCabe-Thiele often has an annoying scaling problem. Most distillations involve high purities; most steps are at the ends of the equilibrium curve and expanded diagrams of their ends must be prepared before any degree of accuracy is obtained. The numerical plate-to-plate method is unaffected by concentration ranges.

EASE OF REPEATED RUNS

A computer based plate-to-plate analysis can be repeated as often as required, and from the design point of view, very many analyses are necessary before a good design is produced.

After two cases are drawn out on a McCabe-Thiele diagram it looks hopelessly confused, unless the technique involves tracing paper and copy machines!

A FOUNDATION FOR GENERAL DISTILLATION DESIGN METHODS

McCabe-Thiele cannot be developed further to allow for pressure drop, enthalpy balances, or multi-component systems. The graphical method for enthalpy balance, the Ponchon-Savarait method, is useful only as an intellectual exercise. The numerical method is the foundation of all advanced design methods - and even if in some cases the equations are not solved with a plate-to-plate procedure, the principles are the same.

The plate-to-plate binary method also serves as an excellent first example of CAD in distillation. It is not too complex a problem, yet all the features of repeated trials, location of optima, and judgement of profiles are involved.

McCabe-Thiele is a good introduction to fractionation. It should remain as the first stage of our teaching of the subject, but it is only an introduction. In no way is McCabe-Thiele the way to design distillation systems.

3.2 NON-IDEALITY IN LIQUID SYSTEMS

How do the non-idealities discussed in Chapter 2 affect the design of distillation columns? If we consider this question by discussing binary distillation then at least we are dealing with a simple system and it is therefore easier to understand the effects.

For ideal systems, $\gamma = 1.0$ and for any binary system, one component always has a higher vapour pressure than the other. Hence for all liquid concentrations, one component will always have a higher vapour concentration than the liquid in equilibrium with it. This means that given enough plates, separations of any degree of purity are possible. For example, water and heavy water can be separated with 500-1000 plates to provide over 95% purity.

If γ is dependent on concentration, as is the case with non-ideal systems, then this simple rule does not hold, and conditions can sometimes exist where the ratio of γ exactly counterbalances the saturated vapour pressure ratios of the two components. In this case there is then no difference in composition between the liquid and vapour phases and no amount of distillation can cause concentrations to change. The system is an azeotropic system and this particular composition is the azeotrope.

For a non-ideal system, the relative volatility at any particular concentration is given by

$$\alpha = \frac{P_1\gamma_1}{P_2\gamma_2} = \frac{y_1/x_1}{y_2/x_2} \qquad (3.10)$$

For an azeotropic composition $x = y$ and $x = y$, $\alpha = 1.0$, and so

$$\frac{P_1}{P_2} = \frac{\gamma_2}{\gamma_1} \qquad (3.11)$$

This situation can occur quite frequently. Whenever the activity coefficient at infinite dilution is greater than the ratio of the saturated vapour pressures of the pure components, it can be expected that an azeotropic exists. An azeotrope is not a peculiar property of a mixture, but simply a condition that results from employing the normal equilibrium equations.

For a binary distillation the presence of an azeotrope means that composition changes in the column cannot cross the azeotropic composition but can only asymtotically approach its value. With a feed above the azeotrope the distillation will remain above the azeotrope. Systems below keep below.

To obtain concentrations on the other side of the azeotrope, methods must be found for changing the relationship given by Equation (3.11). This matter will be handled in Chapter 6.

The Gibbs-Duhem relationship shows that there is a thermodynamic relationship between γ_1 and γ_2. When γ_1 is above unity at low concentrations, γ_2 is above unity at low concentrations; similarly, when the non-ideal forces are attractive, both values are less than 1.0.

For systems with γ above unity, this has the effect of making it easier than 'normal' to remove lights from the heavy component to obtain a high purity bottom product, but correspondingly more difficult to remove the heavies from the lights to produce a high purity light product. The reverse is true when non-idealities involve values less that 1.0. Hence the error introduced by assuming ideality will be dependent on the concentration ranges chosen at each end of the distillation. If the products are equally pure, it may be that an assumption of ideality will give approximately the same number of plates as the non-ideal analysis, but a misplaced feed position. If one end is to be much purer than the other, ideality assumptions will produce too many or too few plates depending on which end was to have the high purity and the form of the non-ideality.

3.2a TWO LIQUID PHASE SYSTEMS

When liquid non-idealities become sufficiently extreme, the exerted partial pressures at low concentrations can be equal to exerted partial pressures at a much higher concentration where there is a much lower activity coefficient. When two compositions exist which exert the same partial pressures, then the stable condition for any mixture that can separate into these two concentrations are these concentrations, and they co-exist as two liquid phases.

The thermodynamic condition for equilibrium is that the Gibbs Free Energy for the system is at a minimum. Hence for a liquid/vapour system at equilibrium, the plot of Gibbs Free Energy vs composition must be concave, as is shown by curve 1 of Figure 3.7 for a binary system. For systems that are highly non-ideal, this Gibbs Free Energy curve does not remain concave, as shown by curve 2 of Figure 3.7. In such a situation the Gibbs Energy can be further reduced if the mixture separates into 2 phases of concentrations A and B, and this therefore occurs. Since the system is in equilibrium this also means that the liquids A and B must have identical vapour pressures. This is so because the high non-ideality gives the low concentration component in A the same vapour pressure as it has at the higher composition B, and similarly for the second component. The vapour/liquid equilibrium associated with the two-phase region is shown in Figure 3.8. The horizontal part of this figure represents the two-phase region and in reality the liquid phase between A and B consists of 2 separate phases of concentrations A and B in such proportions that the resulting mixture composition is x. Over this section the vapour has constant composition, which results in an azeotropic situation where separation into pure components cannot be achieved by distillation.

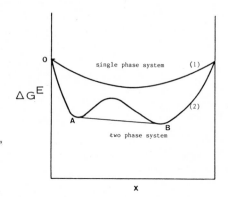

FIGURE 3.7
GIBBS FREE ENERGY PLOTS FOR MISCIBLE AND PARTLY-MISCIBLE SYSTEMS

The identification of the existence of two phases is possible by determining whether the Gibbs Free Energy curve becomes non-convex. This is represented by the situation in which the second derivative becomes negative:

$$\frac{d^2 G^E}{dx^2} < 0 \qquad (3.12)$$

The points where this occurs identify the compositions of the two phases.

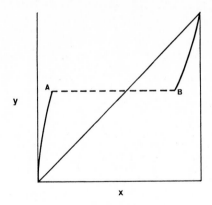

FIGURE 3.8
X/Y DIAGRAM FOR A SYSTEM WITH AN IMMISCIBLE REGION

Since the Gibbs Free Energy is related to the activity coefficients:-

$$\frac{G^E}{RT} = \sum x_i \ \ell n \ (\gamma_i)$$ (3.13)

and the VLE model relates activity coefficients to concentrations, it is possible to develop the required relationship between G^E and x and differentiate this to test for conditions given by Equation (3.12).If two phases are shown to exist, the vapour composition can be determined by the normal VLE model at point A or B.

Only the NRTL and UNIQUAC VLE models are capable of representing two liquid phases, and so such analyses can only be carried out when these are taken as the VLE models. Distillation computer programs do exist that make these two-phase checks, indicate where two-phase regions occur and make the appropriate vapour composition calculations to enable the column calculation to procede. Such programs are rather the exception; they are expensive in computer time and computationally rather tricky.

It is possible to obtain useful information in the calculation of distillations in which two-phase regions occur, using methods based on the single-phase approach. The argument is as follows.

Figure 3.9 shows the x/y diagram for a system with two liquid phases and the fits that are possible using Wilson and UNIQUAC (or NRTL) models. If the UNIQUAC model is used with a normal distillation program, then over the two-phase region, completely unreasonable vapour compositions result, because over the immiscible region alternative program logic must be provided since this region is physically infeasible. However, the Wilson equation, which because of the mathematical form is incapable of the inversions of the more complicated equations, can produce a reasonable fit assuming that two-phase regions do not occur. This fit can then be used, ingoring the fact that the system is immiscible, and reasonable results, including the prediction of the azeotrope, will be obtained. Such a calculation will not, however, give any information about the existence of a two-phase region. This is sometimes important, when, for instance, a two-phase condensate is separated, one phase being returned to the column, and the second phase being removed from the system.

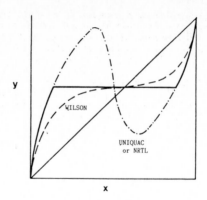

FIGURE 3.9
UNIFAC AND WILSON FITS TO X/Y DATA FOR A PARTLY-MISCIBLE SYSTEM

3.3 DESIGN AND SIMULATION MODES

Consider we require a distillation tower to produce 10,000 tn/y of product with a specification of 1% heavies in the top product. The bottoms product from the tower should have less than 1% top product in it.

Here we have a 'design' distillation problem. To satisfy this given requirement we calculate the number of plates and diameter of the column. An alternative problem is to say we have a 12-plate, 1.0m diameter column available, and a feed of a given composition, what would the performance of this tower be with this feed?

This is a 'simulation' problem. A simulation problem can   easily   be recognised as being equivalent to the possibilities open to an experimenter in the laboratory or plant.

Depending on the problem the chemical engineer has to solve, one or other of the two calculation modes is the more suitable.

In the case of the binary distillation it is possible to have computer programs that function in either one mode or the other (or an option within the same program). The two modes require the equations to be solved in different ways, with different variables being taken as the dependent and independent variables.

As we will see later, this option is open only to binary distillation calcula- tions.   For multicomponent systems the equations cannot be solved reliably in the 'design' mode, thus multicomponent programs are written to perform only in the 'simulation' mode.

3.4 UNDERSTANDING THE PROFILE

In distillation design it is important to have an understanding of the profile of the compositions and temperature vs plates (or column height), which are presented by a thorough analysis. The shape of the concentration profile indi- cates whether the reflux is too high or too low and whether the feed point is correctly located. The temperature profile indicates whether sensing and con- trol problems are likely to occur.

Figures 3.10 - 3.13 give a series of possible concentration profiles that are
possible from a binary distillation. Figure 3.10 shows a good profile assuming
that equally pure top and bottom products are required from the column. Figure
3.11 is the profile that is obtained when the reflux ratio is too low for the
required separation. It is equivalent to the 'minimum reflux' condition, or
the 'pinch' situation of the McCabe-Thiele analysis, and the long section in
the middle where virtually no separation occurs is a complete waste of investment in column height. Figure 3.12 indicates that the column has an inadequate
number of plates. The reflux rate is quite reasonable, as shown by the
inflection, but the required product specifications are not achieved.

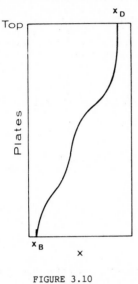

FIGURE 3.10
CORRECT PROFILE

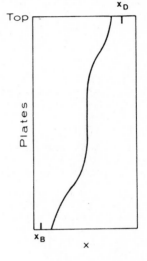

FIGURE 3.11
TOO LOW REFLUX RATIO

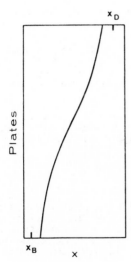

FIGURE 3.12
TOO FEW PLATES

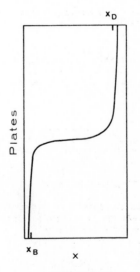

FIGURE 3.13
TOO HIGH REFLUX

Figure 3.13 with the single S curve is a typical profile for a binary distillation, but without the inflection it means that there is enough reflux to prevent any indication of a 'pinch' forming in the column. This is a classic example of a good profile, but no attempt has been made to keep the reflux ratio (and hence energy cost) low. It is unlikely that such a profile would represent an economic design.

Figure 3.14 indicates an incorrectly positioned feed. Above the feed, virtually no separation is taking place; below the the feed, separation is fast. By raising the feed a few plates, the non-performing plates will be exchanged for the good performance shown just below the feed plate. If the feed is moved too high, pinch conditions will be exhibited below the feed and separation will be good above the feed. The 'optimum feed point' is that point that gives the minimum number of plates for a given separation. Hence the optimum feed point occurs when the slopes of the profiles below and above the feed are the same. The profile should appear as a continuous curve (see Figure 3.10).

Figure 3.15 shows the effect on the profile when the incorrect quantity of top product is removed from the column. By 'incorrect' is meant that more is removed than the quantity of light component fed to the column. Hence the distillate must contain considerable proportions of the heavy component, not because there are not sufficient plates available to separate it, but simply because of mass balance: there is no other component to take out to satisfy the distillate rate demanded from the column. In such a situation, the bottom product purity will be higher than required, and the profile would have 'shifted' upwards to satisfy the mass balance.

Of course, these 4 profile characteristics indicating error situations need not occur separately - they will often be compounded. This makes diagnosis much more difficult, but the first step is to recognise the individual symptoms.

As a simple example of interaction, if the wrong distillate rate were specified for a separation of an otherwise correctly designed tower, the 'shifted' profile would exhibit features of an incorrect feed point, and possibly inadequate plates, caused purely by the shift.

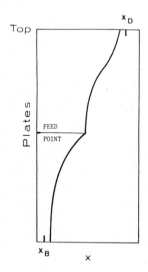

FIGURE 3.14
INCORRECT FEED POINT LOCATION

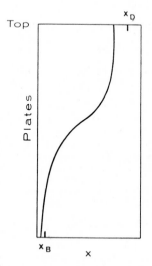

FIGURE 3.15
INCORRECT MASS BALANCE

Though binary profiles can exhibit all these symptoms of inadequate design, many are never evident in a profile calculated with a 'design' program. Binary design programs proceed with an overall mass balance, thus a distillate flow incompatible with the feed flow is impossible. The program actually places the feed at the optimum feed point during the calculation, and the top and bottom product specifications are the starting and stopping points of the calculation, and so the correct number of plates is always produced. For a design program, therefore, the only remaining symptom to be recognised is incorrect reflux ratio (Figures 3.11 and 3.13).

Results from a binary simulation program can exhibit all these symptoms discussed in this section, although many can be avoided by a correct set of input data resulting from a hand mass-balance calculation which ensures that correct distillate rates are specified.

In the case of multicomponent systems, as we shall see later, it is often not possible to specify the distillate rate from a mass balance before the calculation is done. For multicomponent systems therefore the profile becomes the only way of recognising unsuitable input data and of making the required modifications. When it comes to multicomponent design, understanding the profile is essential. The best time to start learning about profiles is during the study of the simpler binary systems.

3.5 THE FEED

There are two important characteristics of the feed that are important for design: the feed position in the column and the heat content of the feed.

The optimum feed point position has been discussed in the previous section and the point was made that the condition for optimum location of the feed is that a smooth profile exists with no discontinuity at the feed point. For a binary system this corresponds to the feed being added to the plate with the same composition as the feed: the liquid composition if the feed is a liquid at its boiling point, and the vapour composition if the feed is a vapour at its dew point. This simple rule applies to binary systems, but not to multicomponent, one reason being that for a multi-component system there is no plate that corresponds to the feed composition for all components. Such a rule predicts N-1 different feed points for an N component system.

The heat content of the feed (or the feed 'condition') is of importance in the calculation of the number of plates. The feed condition determines the flows in the distillation column.

Assuming that we have a feed of rate F consisting of part liquid and part vapour (fraction q liquid and 1-q vapour). Since the stream is well mixed, equilibrium exists and the gas is at its dew point and the liquid at its boiling point. Figure 3.16 indicates the resulting flows within the column.

If the feed is a subcooled liquid, some vapour in the column will be condensed at the feed plate to heat the feed to its boiling point. If the feed is a superheated gas, then liquid from above the feed will be vaporised to cool the gas to the dewpoint. Hence the thermal condition of the feed will alter the L/V ratios at different parts of the column and, as we have shown, this ratio is critical to the separation and calculation of the number of plates in the column.

Defining the q value for the feed is clearly a method of compensating for this problem for mixed vapour and liquid feeds. It can also be used for cases where superheated or subcooled feeds are involved. By converting the degree of superheating or subcooling into an equivalent condensation or vaporisation potential of the feed mixture, this feed enthalpy can be converted to a q value: negative for superheated vapour and above 1 for a subcooled liquid.

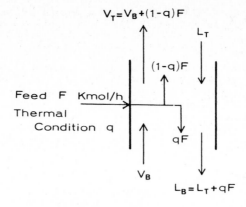

Superheated vapour $q < 0$
Vapour at dew point $q = 0$
Liquid at boiling point $q = 1$
Liquid below boiling point $q > 1$

FIGURE 3.16
EFFECT OF THERMAL CONDITION OF FEED

It can be done by defining q as the heat content of the feed, defined by:

$$q = \frac{\text{Heat to bring 1 mol of feed to vapour at feed plate temperature}}{\text{Molar latent heat of vaporisation of feed}} \quad (3.14)$$

The use of q is a simple and convenient way of relating the heat content of the feed to the resulting flows in the column for the case of a constant molal overflow calculation.

With distillation calculation procedures that take into account enthalpy changes, the use of q is unnecessary because procedures exist for calculating the enthalpy of the feed, and this simply goes into the enthalpy balance at the feed plate.

3.6 PRESSURE DROP

Whatever mass transfer device is used in the column to provide mass transfer for the separation going on, it causes a pressure drop to develop across the column. How important is it to consider this pressure drop?

Distillation is of importance only because separations are difficult: if the constituents separated easily, a single plate, or flash, may suffice. If this were not adequate, a few plates would suffice. A few plates causes no problems of pressure drop and it is not until the separation becomes difficult, requiring many plates, that that we have a distillation 'problem' worthy of deeper study. It is precisely in connection with these difficult problems that pressure drop effects become significant.

Firstly, the pressure drop, which can be as much as 1 bar for a 100 plate column, affects the boiling point on the plates. The bottom of the column now runs hotter than a calculation neglecting pressure drop would predict.

This can have serious consequences if the material in the boiler is heat-sensitive. It may require sub-atmospheric pressures at the head of the column, or an oversized column, or special low pressure drop packing - purely to overcome the thermal decomposition problem introduced by the column pressure drop.

Secondly, the pressure drop is very sensitive to throughput. A 10% reduction in throughput will reduce the pressure drop by 20% and temperatures will change accordingly. This means that in a column with appreciable pressure drop, a temperature within the column - particularly near the bottom - cannot be satisfactorily used to sense the position of the concentration profile for control purposes. Control becomes more difficult in a column with pressure drop.

Thirdly, the high pressures and temperatures at the bottom of the column affect the VLE. It is generally true that the higher the system pressure, the closer together the boiling points are for any two components. Hence the higher the pressure, the more plates are needed to achieve a given separation - and this increase in plates produces an even higher pressure drop.

The effects of pressure drop are detrimental. Pressure drop is a necessary evil when high mass transfer performance is demanded to achieve low capital costs. One of the major factors in choosing column internals, or developing new internals, is to consider pressure drop per theoretical plate at design loadings. Pressure drop must be calculable for the internals, and must be allowed for in the calculation of separations involving many plates.

3.6a THE INCLUSION OF PRESSURE DROP IN BINARY DESIGN CALCULATIONS

The binary design calculation starts with a known bottom composition from the overall mass balance and calculates plate-to-plate until the top composition is reached. Now that we introduce pressure drop we no longer know the bottom conditions completely because top pressures are usually specified and the bottom pressure is the specified top pressure plus column pressure drop.

Hence, modifing an existing binary calculation to include pressure drop involves introducing iteration, first to calculate the number of plates without pressure drop, then calculate the pressure drop, and then repeat the design with the new bottom pressure to convergence. Convergence is achieved when the number of plates calculated by successive iterations is constant.

Convergence may take more than one iteration, because the introduction of pressure drop decreases relative volatilities and increases the number of plates. Generally a few iterations give a satisfactory solution, but sometimes problems occur. For example:

Let the first pass though the calculation require 50 plates, assuming zero pressure drop for a particular problem. A pressure drop per plate of 0.02 bar is specified. This means that the second iteration uses a bottom pressure of 2bar. In this example, at 2 bars the relative volatility is much worse and it now requires 110 plates for the separation. During this calculation 0.02 bar is removed per plate up the column, and so by the 100th plate the total pressure is

$$2.0 - 100 \times 0.02 = 0.0 \text{ bar}$$

The distillation calculation fails because the pressure is zero.

This situation has been described in detail to show the type of numerical problems that occur in distillation and how a minor modification to 'correct for pressure drop' by carrying out iterations can have far reaching effects.

However, a little thought can overcome the problem. If we calculate from the top down, then in a single calculation we can determine the number of plates in

the column; no iterations, no instability as long as the column top pressure
is specified and not the boiler pressure. The procedure must simply work
through the equations in reverse using a dew point calculation in place of a
bubble point calculation.

The BINARY program detailed in the Appendix and Exercises works from bottom to
top and so does have a weakness in that negative pressure drops can occur. This
presents a challenge to the enterprising reader to develop a better 'top-down'
method.

It is generally advisable first to investigate a new distillation problem
neglecting pressure drop, and only after an initial appreciation of the problem include the plate pressure drop.

3.7 CONTROLABILITY

Part of the design of a distillation columns is the specification of the control system to be used. This in turn requires an assessment to be made of the
suitability of the temperature profile for control purposes. The first requirement is that there is a measurable profile, where a point can be found at
which the temperature responds to the concentrations in the column with sufficient sensitivity for control but not in so sensitive or non-linear a fashion
that it would cause instabilities in any control loop. Figure 3.17 shows a
typical temperature profile of a distillation very suitable for temperature
control. The selection of a suitable position for the temperature sensor is
part of the designer's duties, and suitable points are marked with an X on Figure 3.17. At these points there would be a steady signal, approximately
linear over a wide range.

Position Y is much less suitable for control because it is at a point so sensitive in the column that the temperature falls from the upper to the lower value
with only a very small change in the position of the profile. Such a signal
cannot be used for control purposes because it is not linear enough to provide
a smooth control signal for, for example, the column reboiler.

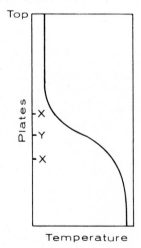

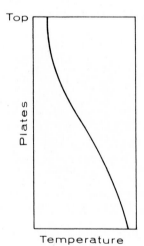

FIGURE 3.17
TEMPERATURE PROFILE SUITABLE
FOR CONTROL

FIGURE 3.18
TEMPERATURE PROFILE UNSUITABLE
FOR CONTROL

Figure 3.18 shows the profile obtained for a difficult separation requiring many plates and in a column with an appreciable pressure drop.

In such a situation the profile is governed more by the column loading and resulting pressure drop than by composition. Hence, any column throughput change will completely mask any temperature changes due to composition. Simple temperature sensors cannot be used for controlling such a column. If the situation is not too serious, it may be possible for a sensor near the top of the column to be sufficiently pressure-drop-independent to be used.

In theory it is possible to measure both the total pressure and the temperature at a suitable point in the column and correct the temperature signal by the pressure to get a true composition dependent temperature.

Given reliable pressure measurement, conversion and temperature correction equipment, the method must work. In the author's experience, it can be said that, given unreliable equipment, the method does not work!

3.8 ROBUSTNESS TO PROCESS FLUCTUATIONS

The calculated number of theoretical plates to achieve the required output assumes perfectly steady state conditions. Whenever fluctuations occur, then with that number of plates the products will be off specification.

Fluctuations are always occurring in a real process. Firstly, control systems operate by allowing deviations from the set point to take corrective action; there are always small fluctuations to enable the control system to operate. The size of these fluctuations depends on the sensitivity of the sensor, but there needs to be more than one plate in the column than calculated to allow these fluctuations to occur without producing off-spec product.

Secondly, feed composition fluctuations can occur in a random fashion from disturbances produced in up-stream processes. Again, the number of plates supplied needs to be greater than that calculated in order to enable a fluctuating feed to be handled without producing problems in product purity. Given a range over which fluctuation can occur, the extra number of plates can be calculated by further distillation calculations.

Thirdly, throughput in the column causes changes in the temperature profile that must be appreciated and allowed for in the design. A reduction in throughput of 50% will produce a reduction in pressure drop to one quarter. Hence, the pressure drop will virtually disappear, and the temperature profile will be different in that the pressure in the whole column will now be equal to the column top pressure. This will make a few degrees difference in the temperature in the bottom of the column. A temperature sensing point in the column will hold a particular point in the tower at a fixed temperature. Hence, when throughputs reduce, and the control system still maintains the same temperature at the same point in the column, then although the temperature is controlled, the composition profile will move up the column. The temperature drop due to the lower pressure will be compensated for by more heavy components in the mixture. In a 'robust' design this should be compensated for by adding extra plates at the top of the column so that the product will still remain in specifications when rates are reduced. It is not a good design if the setpoint of the controller must be changed whenever rates are changed.

The number of extra plates can be calculated by running the design program with and without pressure drop and noting, at the selected temperature point, what shift in profile occurs, and how many extra plates this is equivalent to.

3.9 FLEXIBILITY TO ALTERNATE FUTURES

In some processes, distinct alternative futures are possible. An oil refinery may be fed with a Middle East crude, or possibly a Venezuelan; a chlorination plant may require its major product to be the di-chlorinated product, or possibly equal amounts of di and tri; 10,000 tn/y of product may be required, or possibly 15,000.

If there are grounds for considering any of these possibilities seriously, any distillation column design for that plant should also consider the implications if these possibilities were to occur.

In addition to studying the design case, one should also study the performance of the proposed equipment under the alternative conditions. This is most convenient to do with a computer program that uses simulation mode as opposed to the design mode, so that the behavior of the proposed column can be simulated easily. Studies can, however be made – albeit less conveniently – using a design program.

It may become obvious from such studies that the alternative would require only minor design changes to allow the alternative future to be handled by the same design. A feed point change may be the only modification required, and this may be conveniently included in the specification by simply requesting two feed points for the column. A slightly higher column may be required, and it may be considered worthwhile to specify this in the design. Even if no changes to the design result from the study, the engineer at least knows what would have to be done, and this may affect the layout of equipment.

3.10 SAFETY FACTORS

Engineering design is carried out using design procedures that are considered to best represent the process under study, and these design procedures require numerical values for their parameters, which are again supplied as 'best' or most-likely values. However, both the method and the parameters are not precise and their inaccuracies – given by standard deviations of the input data and parameters – results in an error in the final design. In fact the resulting design given by the design method has a 50% chance of being too big, and a 50% chance of being too small. In other words, there is a 50% chance that the column will not meet design throughput.

The philosophy behind the choice of appropriate confidence level is a complex optimisation problem. It is necessary to build something that is most probably going to be too big in order to ensure that some future capacity – that may never be needed – can be attained. This optimisation is a balance between loss of future income, which is related to accuracies of sales forecasts, growth rates and profit margins, and cost of providing marginal increases in capacity. Numerous studies have shown that, economically speaking, surprisingly low confidence levels of achieving the rated output should be taken. In cases of low profit margins, safety factors should be negative!

This is a direct contradiction to most engineering design philosophy which – purely for political reasons – demands a high certainty that the plant will achieve its rated output. If politics demands it, or contracts guarantee it, the 95% confidence levels must be taken even though it may not make economic sense.

A further problem is how the safety factor should be divided so that the overall plant has the right confidence level. A chain of 10 items, each with a 95% confidence level, will only give

$$0.95^{10} = 0.60$$

confidence level for the whole plant. Each item should not have the same confidence level. Small,inexpensive, well understood items (e.g. pumps) should be designed to 99% levels, leaving the large,expensive, less understood items(e.g. distillation columns) to carry lower levels.

When the problem is approached theoretically, it can be shown that when designs become very uncertain, it is preferable to have a very low safety factor, and be prepared to modify or extend the item if it proves to be too small (which will anyway have a likelihood of less than 50%). Again such a logical policy can only be carried out if it is politically acceptable to modify the plant after start-up if the equipment proves to be too small. Methods do exist for such analyses, but it is unusual for them to be applied in practice. Generally, one needs a 95% probability that design throughput will be achieved. Hence, we must specify more plates than our calculation suggests to raise the probability from 50% to 95%.

Statistically speaking, we need to add 2 standard deviations in the accuracy associated with the number of plates, which in turn is related to the accuracy of the VLE data. The standard deviation for the number of plates could be determined from the error on the VLE data using Monte-Carlo simulation - but this is not general practice. More usually, one adds a percentage on to the column, the magnitude of which depends on the reliability of the VLE used. 10% may be considered adequate when reliable VLE data are used. Hence, a final design is made up of:

Plates above feed:

$$N_T \quad \text{as calculated above the feed}$$
$$+ \; 0.1N_T \quad \text{10\% for parameter accuracy}$$
$$+ \quad 3 \quad \text{plates for robustness}$$

Plates below feed:

$$N_B \quad \text{as calculated above the feed}$$
$$+ \; 0.1N_B \quad \text{10\% for parameter accuracy}$$

$$\overline{1.1(N_T + N_B) \; + \; 3} \quad = \quad \text{total number of plates to specify}$$

To this would be added any extra thought necessary to improve the flexibility as described in Section 3.9 .

3.11 AN EXAMPLE

A column is required to recover methanol from a mixture of 30 mol% methanol and 70 mol% water. The recovered methanol should have a purity of 98 mol% and the discharged water should contain less than 1000 ppm(wt) (5.6×10^{-2} mol%) methanol. The feed rate is 220 kmol/h.

A plate column is to be used with a plate efficiency of 85%. The design vapour velocity in the column can be considered to be 0.37 m/s. Under these conditions the plate pressure drop will be 0.01 bar/plate.

(1) Propose a suitable column design

(2) How should the column be controlled?

(3) If the feed concentration changes to 50 mol% methanol, how would the proposed column operate?

Cost Data: Steam costs: 20 $/t
             Column capital costs: assume 6000 $/M^2 of plate.
          (Chapter 7 considers capital costs in much more detail)

The first rule of CAD is to carry out some approximate hand calculations to understand the order of magnitude of the problem and at the same time to prepare the necessary input to the computer program.

Firstly, a mass balance over the column is needed, and this can be constructed directly from the data defining the problem. This is given on Figure 3.19 .

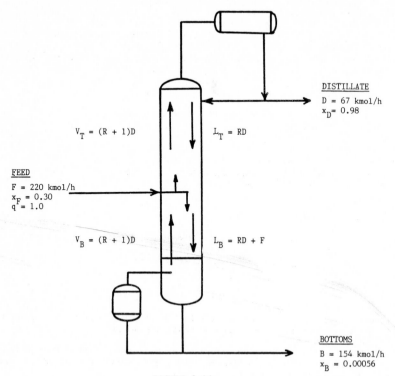

FIGURE 3.19
MASS FLOWS OVER THE COLUMN FOR THE EXAMPLE

Following this, we can carry out a few simple calculations to determine the magnitude of the problem:

(1)   How big is our project? At 220 kmol/h feed, with a mean molecular weight of about 20, the yearly capacity is about

$$\frac{200 \times 20 \times 8000}{1000} = 35,000 \text{ tn/y}$$

This is quite a large unit.

(2) What degree of separation difficulty will there be?   If we assume ideality, we can calculate the relative volatility from the vapour pressures at 100 Deg C:

$$P_{MeOH} = \exp(11.855 - 3571.47/(373 - 36.5) = 3.5$$

$$P_{H_2O} = 1.0$$

Hence, $\alpha = 3.5$

That is quite an easy separation. For example, 10 plates with a 3:1 reflux ratio may be suitable.

(3) What column diameters are we considering? The vapour flow is derived from
Figure 3.19 with a 3:1 reflux ratio and has units kmol/h. Column loading is
best appreciated in m/s gas velocity. Hence, we need to convert V to m^3/s at 100
Deg. C.

$$V \text{ (kmol/h)} \times 22.4 \times \frac{373}{273} \times \frac{1}{3600}$$

$$= 8.5 \times 10^{-3} \text{ V } m^3/s$$

For a design velocity of 0.37 m/s, this gives the column cross-section as

$$\frac{8.5 \times 10^{-3} \text{ V}}{0.37} \quad = \quad 2.26 \times 10^{-2} \text{ V } m^2$$

Taking our rough estimate of reflux ratio of 3:1, from Figure 3.19:

$$V = 67 \times (3 + 1) \quad = \quad 268 \text{ kmol/h}$$

Hence, the cross-section area will be

$$268 \times 2.26 \times 10^{-2} = 6.1 \text{ } m^2$$

which corresponds to a column diameter of

$$\sqrt{\frac{4 \times 6.1}{\pi}} = 2.8 \text{ m}$$

We can carry out this rough calculation further to investigate the investment
costs and energy costs to see where the major costs lie.

In this example, the marginal equipment costs have been correlated with the total area of plates. For our estimated column with a reflux ratio of 3:1 and 10
plates, the 'capital' required is

$$6.1 \times 10 \times 6000 \quad = \quad 366 \times 10^3 \text{ } \$$$

and the steam costs, assuming no losses, are equivalent to the quantity of
vapour generated in the boiler, since the bottom component is essentially also
steam:

$$1 \text{ kmol/h steam costs } \frac{20 \times 18 \times 8000}{1000} \quad = \quad 2880 \text{ } \$/y$$

Hence, for our estimate of vapour flow of 268 kmol/h the resulting annual cost
will be

$$2880 \times 268 \quad = \quad 775 \times 10^3 \text{ } \$/y$$

Our first impression is therefore that steam costs are more significant than
capital costs.

The first step in the detailed solution of this example is to locate an
'optimum' reflux ratio. Given equipment costs (on a marginal basis) and energy costs, it is possible to locate the optimal design conditions for this particular situation. The general rule to choose reflux ratios between
1.2 and 1.5 times the minimum reflux ratio is a good starting point. A

closer specification depends on a balance of the particular investment costs
and energy costs. The optimum will be very different if we are considering a
stainless steel tower and low energy costs or a mild steel tower with expensive
energy.

The computer methods cannot deliver the minimum reflux as simply as the McCabe-
Thiele method, but it is relatively simple to run off a set of design cases
with decreasing reflux ratio. When no results can be obtained because the maximum allowed number of plates set (e.g,.300) has been exceeded, the minimum reflux ratio can be determined by extrapolation, as shown by Figure 3.20.
By plotting 1/R vs. 1/N, the point where 1/N is zero gives the minimum reflux
ratio.

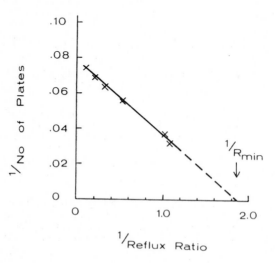

FIGURE 3.20
LOCATION OF THE MINIMUM REFLUX RATIO FOR THE METHANOL/WATER EXAMPLE

Once a set of runs at different reflux ratios has been made, then we are in a
position to choose an optimum design for the particular set of cost conditions
applying to this problem.

This study is best done in tabular form, calculating for each design the corresponding capital and energy costs and combining them together to form the total
annual cost. The best design can then be selected from this table.

It is more convenient, and less susceptible to error, to determine the
algebraic expressions that determine the appropriate costs directly from the
variables reflux ratio and number of plates. These expressions are easy to develop:

In the same way as the development of the expression for capital cost, we can
say

Marginal capital costs $= 2.26 \times 10^{-2} \times 6000 \times D(R+1)N$ $

If we assume that the annual cost caused by the marginal investment cost is 25% of the capital cost (e.g. 15% return on investment and 10% depreciation) then :

$$\text{Annual cost from column capital} = \frac{2.26 \times 10^{-2} \times 6000 \; D(R+1)N}{4}$$

$$= 33.9 \; D(R+1)N \quad \$/y$$

The annual steam cost can be derived from our price per kmol of steam:

$$\text{Annual steam costs} = 2880 \times V = 2880 \times D(R+1) \quad \$/y$$

For our production rate, D is constant at 67 kmol/h.

We are now in a position to tabulate simulated results of total costs, and also individual capital and energy costs for various reflux ratios (see Table 3.1).

This Table is best constructed as the study progresses. After the first run the contributions of the capital and energy are entered in the table. This immediately indicates whether or not the next run should be at a greater reflux ratio or a smaller one. The search for the optimum is not random but based on the information contained in the Table.

As the work continues, and the table gets longer, it becomes obvious that the possible improvements in total cost are becoming very small. When these differences are no longer significant, the optimization phase of the design is completed. Note that our criterion is not to locate the reflux ratio within 1% of its optimum value but rather to locate the annual costs to within 1% of its minimum. This first criterion is much the more difficult because of the flatness of the optimum, as the table clearly shows.

From such a table, a design reflux ratio of 1:1.0 would seem to be quite reasonable. There is no point in carrying out more accurate work to decide whether 1.01 or 0.99 would be better, since the difference in annual savings would be less than 100 $/y, which in view of all the uncertainties in the system is not different from zero.

TABLE 3.1

EXAMPLE - LOCATION OF OPTIMUM REFLUX RATIO

			(D = 67 kmol/h)	
Reflux Ratio	No of Plates	Annual Capital Costs	Annual Steam Costs	Total Annual Costs
R	N	$33.9D(R+1)N$ (a) ($ x10^3)	$2880(R+1)D$ (b) ($ x 10^3)	(a) + (b) ($ x 10^3)
5	14.7	200.3	1157.7	1358.0
3	16.4	149.0	771.8	920.8
2	18.5	126.0	578.9	704.9
1.2	24.0	112.0	424.5	536.5
1.1	26.8	123.0	405.2	528.2
1.0	28.8	130.8	385.9	516.7
0.95	31.8	140.8	376.3	517.1

Having chosen the reflux ratio, the temperature profile can be investigated to choose a suitable control point. Table 3.2a gives the column profiles for the design condition.

TABLE 3. 2
COMPARISON OF TEMPERATURE AND COMPOSITION PROFILES

TABLE 3.2a FOR NORMAL OPERATION				TABLE 3.2b AT LOW THROUGHPUTS			
Plate No	Temp K	Press bar	X	Plate No	Temp K	Press bar	X
2	379.0	1.26	0.001	2	373.0	1.02	0.001
3	378.4	1.25	0.002	3	372.7	1.02	0.002
4	377.2	1.24	0.007	4	371.5	1.02	0.008
5	374.0	1.23	0.024	5	368.3	1.02	0.027
6	368.2	1.22	0.069	6	362.3	1.02	0.077
7	362.1	1.21	0.148	7	356.5	1.02	0.162
8	358.2	1.20	0.225	8	353.2	1.02	0.238
9	356.3	1.19	0.273	9	351.7	1.02	0.282
10	355.3	1.18	0.297	10	350.8	1.02	0.313
11	354.1	1.17	0.331	11	349.5	1.02	0.363
12	352.6	1.16	0.377	12	348.0	1.02	0.423
13	351.0	1.15	0.433	13	346.6	1.02	0.490
14	349.5	1.14	0.495	14	345.3	1.02	0.558
15	348.0	1.13	0.558	15	344.1	1.02	0.622
16	346.6	1.12	0.619	16	343.1	1.02	0.681
17	345.4	1.11	0.676	17	342.2	1.02	0.734
18	344.3	1.10	0.728	18	341.5	1.02	0.781
19	343.3	1.09	0.774	19	340.8	1.02	0.823
20	342.4	1.08	0.816	20	340.2	1.02	0.859
21	341.6	1.07	0.852	21	339.7	1.02	0.892
22	340.8	1.06	0.885	22	339.3	1.02	0.920
23	340.2	1.05	0.914	23	338.9	1.02	0.946
24	339.5	1.04	0.940				

Plate No. 12 with a set point temperature of 370 K would seem to be a good control point in that it would have a suitably measurable linear response to small movements in the column profile. This point is discussed in more detail in chapter 8.

To investigate robustness of this proposed control system to throughput changes a further run can be done with zero pressure drop per plate. This gives the profile reported on Table 3.2b. By inspecting both profiles it is clear that a 'shift' of 1 plate would occur if the set point was left at 370 K. Hence 1 more plate should be supplied above the feed to cover this situation.

The case of the alternative feed composition posed in the problem is a little inconvenient to investigate with the 'design' binary program, but it can be done by running a set of simulations at varying reflux ratios with the new feed until product specifications are achieved with 28 plates - the number chosen for our basic design.

This was found to require a reflux of 0.65:1 - requiring the feed at plate 15. Hence we could specify a second feed point in the column to cover this eventuality.

What would the column capacity be with this new feed? The column capacity is limited by the design vapour velocity – this is D(R+1) which for our design is

$$67 \times (1+1) = 134 \text{ kmol/h}$$

With a reflux ratio of 0.65, this gives a distillate rate of

$$V/(R+1) = 134/1.65 = 81.2 \text{ kmol/h}$$

Since the feed is 0.5 mol fraction methanol, this corresponds to a maximum feed rate for the column with the new feed of 162 kmol/h.

We are now in a position to specify the number of plates required in our column

No of plates above feed:
	16	(basic design)
+	1	for control fluctuations
+	1	for P.D. /temperature shift
+	2	10% design margin
total	20	

No of plates below feed:
	12	(basic design)
+	1	for control fluctuations
+	1	10% design margin
total	14	

$$\text{Total plates required} = 34$$
$$\text{Feed plates} = 14 \text{ and } 17$$
$$\text{Temperature control point plate} = 11$$

3.12   CONCLUDING REMARKS

The opportunity has been taken to demonstrate within this chapter on binary distillation all the factors involved in CAD of distillation columns. Binary distillation is an excellent vehicle with which to demonstrate these principles on a comparatively simple system. These factors are all very important also in multicomponent distillation, but the extra problems of multicomponent distillation make it one degree more difficult, as will be seen in the follow-ing chapter.

3.13 FURTHER READING

King C.J., 1971. Separation Processes, McGraw-Hill.
-good on graphical and analytical methods.

Smith J.M.and Van Ness H.C.,1975. Introduction to Chemical Engineering Thermo-dynamics, McGraw-Hill.
-gives the thermodynamics of partially miscible systems.

3.14 QUESTIONS TO CHAPTER 3

(1) What are the limitations of the McCabe-Thiele method?

(2) How can the minimum reflux be located by McCabe-Thiele, and how by numerical methods?

(3) What is the effect of minor deviations from ideality in the VLE data?

(4) What types of system form azeotropes and why?

(5) Why do some systems form two liquid phases? How can the phenomena be predicted?

(6) What is the difference between Design and Simulation mode calculations?

(7) What information is contained in the profile?

(8) How is pressure drop important?

(9) What is meant by controllability, robustness and flexibility?

(10) Why should more plates be installed than given by the design calculation?

Chapter 4

MULTICOMPONENT DISTILLATION

4.1 INTRODUCTION

We have seen that for a binary distillation we can write an overall mass balance over the column, define top and bottom composition, define distillate rates to achieve these compositions, and calculate the number of plates needed to achieve the separation in a single plate-to-plate pass through the column - our design calculation.

Once we have more than two components, this situation is changed. We can no longer define our top and bottom composition. We now have too many unknowns to define a bottom composition that is achievable in practice. We are therefore not certain of our best distillate rate, since its composition is not known. Since we do not know our end compositions specifically, we have to iterate over the whole column calculation to convergence. The convergence of the calculation is a perpetual difficulty in multicomponent distillation and thus reliable convergence methods must be used. Under no conditions should discontinuities caused by introducing integers into the problem be allowed to interfere with the convergence. Hence the number of plates must not be varied. We are   faced with a 'simulation' mode calculation to simulate the performance of a given piece of equipment. Design mode calculations for multicomponent distillations are not practicable.

In such a situation, how can we start with a good proposal with which to simulate? The normal technique is to use a short-cut analytical method of designing a column to meet a given product specification. This method then gives a proposed design that can be refined by the thorough multicomponent distillation calculation.

4.2 SHORT-CUT METHODS

The short-cut methods can generally only be expected to provide a rough result because the methods are only strictly accurate under two conditions:

(a) that the system is 'ideal', and each component can be represented by a relative volatility which is constant over the whole range of compositions involved,

(b) that the 'keys' involved in the separation are adjacent, i.e when no components have volatilities between the two components to be separated. The idea of adjacent keys is discussed in detail during the discussion of multicomponent profiles later in this chapter.

When these conditions do not apply, it cannot be expected that the short-cut methods will give an accurate result. However,it will still give a result with which to start the simulation study.

Short-cut distillation procedures are based upon the work of Fenske, Underwood and Gilliland. A combination of their work provides a full picture of the relationship between numbers of plates and reflux ratio, and gives an indication of the splits to be expected with the other components when a specific separation is achieved between the two key components.

4.2a THE EQUATIONS

Fenske showed that at total reflux the relationship between the molar ratios of the key components in the top and bottom products could be related to their relative volatility (α) and number of plates N_{min} by the expression:

$$\alpha^{N_{min}} = \frac{\left(\dfrac{x_{LK}}{x_{HK}}\right)_D}{\left(\dfrac{x_{LK}}{x_{HK}}\right)_B} \tag{4.1}$$

Underwood developed a relationship to determine the minimum reflux ratio (or minimal L/V ratio) for a multicomponent system:

$$\left(\frac{L}{V}\right)_{min} = \sum_{i=1}^{n}\left(\frac{x_{D_i}}{\dfrac{\alpha_i - \theta}{\alpha_i}}\right) - 1 \tag{4.2}$$

where θ is a variable determined by the solution of the equation:

$$\sum_{i=1}^{n}\frac{\alpha_i \, x_{F_i}}{\alpha_i - \theta} = 1 - q \tag{4.3}$$

Equation (4.3) relates the feed composition of all components (x) and the feed heat condition q where:

$$q = \frac{\text{Heat to bring 1 mol of feed to vapour at feed plate temperature}}{\text{Molar latent heat of vaporisation of feed}} \tag{4.4}$$

Equation (4.3) requires iterative solution, and this is conveniently and reliably carried out by the simple single dimension secant method. Having evaluated θ , this can be used in equation (4.2) to determine $(L/V)_{min}$ Although the method requires distillate mole fractions of all components, these are strictly not known without carrying out a full multicomponent calculation. However if we make the assumption that the two keys are adjacent and all the compositions are entirely light or heavy then we can, by means of simple mass balance, determine the value of x for all components with sufficient accuracy for this method to be used to provide initial estimates.

Equations(4.1) and (4.2) therefore provide the two asymptotes for the well known number of plates versus reflux ratio relationship. What is now necessary is to complete the curve and so enable the number of plates at different reflux ratios to be determined.

Gilliland plotted the relationship shown in Figure 4.1 for a large number of different systems and showed they all lie on approximately the same curve. This curve is entirely empirical, but it does provide the necessary link between Fenske and Underwood so that an actual operating condition can be predicted. Knowing $(L/D)_{min}$ from Underwood, and taking an operating ratio L/D (usually expressed for convenience as a factor times the minimum reflux)

e.g. $L/D = f \, (L/D)_{min}$

then the X-axis of Figure 4.1 is determined, allowing the corresponding Y-axis value of to be read off the curve. A little algebra then allows N to be determined.

For computer use the empirical curve of Gilliland must be replaced by an equation, and a suitable fit has been published by Eduljee as:

$$\frac{N - N_{min}}{N + 1} = 0.75 - 0.75 \left(\frac{\frac{L}{D} - \left(\frac{L}{D}\right)_{min}}{\left(\frac{L}{D} + 1\right)} \right)^{0.5688} \tag{4.5}$$

Our short-cut method now goes further by also giving a prediction of the expected distribution of the non-key components. This is done again according to the Fenske relationship:

$$\frac{\left(\frac{x_{D_i}}{x_{D,LK}}\right)}{\left(\frac{x_{B_i}}{x_{B,LK}}\right)} = \left(\frac{\alpha}{\alpha_{LK}}\right)^N \tag{4.6}$$

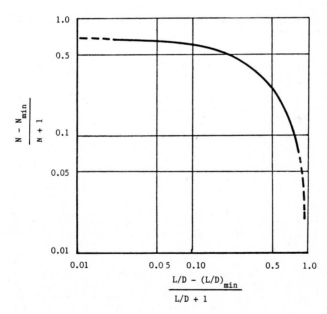

FIGURE 4.1.
GILLILAND EMPIRICAL RELATIONSHIP BETWEEN N AND L/D

Finally, methods are available for predicting the feed point for an ideal case represented by the limitations already put on the method. Kirkbride proposes the following equation for the estimation of the feed point position:

$$N_F = \frac{N}{1 + X}$$

(4.7)

where

$$X = \left\{ \frac{F - D}{D} \cdot \frac{(x_F)_{HK}}{(x_F)_{LK}} \left(\frac{(x_B)_{LK}}{(x_D)_{HK}} \right)^2 \right\}^{0.206}$$

When all these contributions are collected together as one program, this is a quick way of getting some idea of the magnitude of the distillation problem, and gives a good starting value for a simulation study.

Comments on the accuracy of the method are difficult to make. When the relative volatility is not constant, and when intermediate components play an important role in the distillation, then discrepancies are bound to occur as the assumptions of Fenske-Underwood-Gilliland have been violated. However, it is always worth starting a study with this calculation however much assumptions do not hold. When there is difficulty in identifying the 'keys' of the distillation, different 'keys' can be used and the different results compared before a starting point is chosen for the accurate multicomponent simulation.

4.3 MULTICOMPONENT SIMULATION - CONSTANT MOLAL OVERFLOW

The complete multicomponent distillation must handle each plate and determine the equilibrium relationship for each component on the plate. Expanding from the binary description given in Chapter 3, the equation defining the VLE can be written as:

$$y_i = f(x_i)$$

(4.8)

and the basic balance equations now have to be written for every component:

$$V_{N-1} y_{i,N-1} + L_{N+1} x_{i,N+1} = V_N y_{i,N} + L_N x_{i,N}$$

(4.9)

On each plate we have variables x_i, y_i, L, V and T for the n component system, giving 2n + 3 variables. Taking an average problem size of 10 components at 50 plates therefore produces a total of 1150 equations and 1150 variables that have to be solved.

These equations have a special form; apart from the top and bottom plate, the variables are linked only to the plates above and below. The end plates are the exceptions because an overall mass balance must hold over the column. The material leaving must equal that fed to the column, so iteration is required until it is achieved.

Figure 4.2 gives the structural matrix for the equations involved. This has a special sparse structure, and this leads to a number of proposals for its solution.

	xF	9 x (N-1) dependent variables								
		y0	x1	y1	x2	y2	x3	y3	xD	xB
Boiler: equilibrium		x								x
mass balance		x	x							
Plate 1: equilibrium			x	x						
mass balance				x	x					
Plate 2: equilibrium					x	x				
(feed) mass balance	x					x	x			
Plate 3: equilibrium							x	x		
mass balance								x	x	
Overall mass balance	x									x	x

FIGURE 4.2.

STRUCTURAL MATRIX FOR THE MULTICOMPONENT DISTILLATION CALCULATION

4.3a CONVERGENCE METHODS

PLATE-TO-PLATE (Lewis-Matheson Methods)

The particular form of the structure matrix suggest that the diagonal nature should be made use of by solving plate-to-plate proceeding from one end of the column to the other. This is simply an extension of the binary method, except that now we will not be in mass balance over the column with a single iteration and thus some iteration procedure must be used to achieve this mass balance. Various methods are possible, from modified forms of repeated substitution to the use of the more mathematically based methods such as Newton-Raphson or multidimensional secant.

The plate-to-plate method, however, has some severe shortcomings in that it is not very 'robust'. It is frequently found during the iteration that the calculation cannot proceed further because some of the fractions become negative or greater than 1 and resulting equations cannot then be evaluated numerically. The equations describing the system do not behave in a stable fashion when mole fractions become negative or greater than 1.

The instability is introduced because the calculations have to be made with non-real end compositions as initial estimate in order to get a response with which to improve these end estimates. These initial guesses can result in erroneous mole fractions developing during the calculation up the column (see Figure 4.3).

The chance of this divergence occurring can be reduced by calculating only part way of the column (e.g. to the feed plate). A second calculation then starts from the top of the column to the feed and we now check our mis-match with the two feed plate compositions, and iterate until convergence on the feed plate composition is achieved. This method does not solve the instability problem but only reduces the chance of it occurring.

Considerably improved reliability is achieved if the plate-to-plate calculation proceeds up the column only to the onset of instability. At this point this is artificially considered to be the total height of the column, and this system is iterated until convergence, so predicting the performance of the smaller column. After convergence, resulting bottom concentrations are taken as initial guess for the full height of column simulation. If an unstable point is again found - now further up the column - the procedure is repeated, getting a better bottom estimate each time, and getting further up the tower.

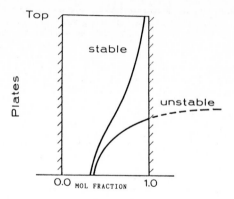

FIGURE   4.3.
STABILITY   PROBLEMS WITH THE PLATE-TO-PLATE CALCULATION.

A variation of this method is to solve the whole column using a very low plate
efficiency. In this way, concentration changes are small, and thus the movement
out of the real region out of the real region can be avoided. Once convergence at a low plate efficiency is achieved, the plate efficiency can be increased, and a new
convergence found. Stepwise increase in plate efficiencies are made in this way
until the appropriate plate efficiency is reached. The mathematical technique
here is that of Davidenko: the solution of a similar but simpler problem on
the way to the solution of the real problem!

These two ways for improving the stability of the plate-to-plate method result
in a considerable increase in computer time, which means that the method develops a variable performance in terms of computer time, depending on whether
these stabilising methods have been used within a particular study.

The plate-to-plate methods require initial estimates of the top and bottom
compositions as a starting point. They have gained a bad reputation for
instability, which will no doubt stay with the method even although probably
unjustified when the the full potential of the Davidenko procedure is adopted.

MATRIX METHODS

The instability of the plate-to-plate calculation can also be overcome if calculations are made over one plate and not over a number of plates in which the
errors accumulate. The matrix methods do not make full use of the diagonal
property of the structural matrix for sequential solution of the equations, but
instead the equations are solved by type and not by plate. These types of solution methods are known as Thiele-Geddes type methods, and there are many variations because of the large number of possible ways of re-arranging the
equations and selecting the independent and dependent variables for their solution.

It is usual in Thiele-Geddes type methods to transform the equations to use
molar flows of each component in vapour and liquid as major variables and to
eliminate mole fractions. The method takes initial values for temperatures,
vapour rates .and concentrations per plate, and from the plate temperatures and
compositions equilibrium K values are calculated, and are these used to solve a
set of linear equations which define the component flows on each plate. (Some
methods start by requesting initial estimates of K values instead of T and x.)
The up-dated component flow can be used to re-calculate V and T per plate and
the procedure iterated to convergence on T and V.

Many variants exist: e.g. details of the solution of the set of linear mass balance equations (a tri-diagonal matrix), and details of the way in which the temperature is re-established. The method of Holland contains refinements in these areas.

The Newton-Raphson (or Linearisation Methods) rely even less on rearranging the equations before solution, and solve for all variables simultaneously using non-linear simultaneous equation-solving methods of the Newton-Raphson type. Again, variants exist in the way in which the sparse matrix is handled and depending on whether the Jacobian is up-dated on each iteration.

The disadvantage of the method is the computer time required to produce the Jacobian (whether by numeric or analytic differentiation) and the storage required for the inversion of the matrices. The Broyden method has been success-fully used to reduce computer time.

The Naphtali-Sandholm method, used in the program DISTIL (see Appendix 4) is a variation of the Newton method.

These matrix methods are based on mathematical solutions to the equation set defining the distillation system, and, opposed to the plate-to-plate and relaxation method, they soon lose physical interpretation. To go further into detail it is necessary to look at the equations more exactly in a purely mathematical sense. Since it is not the intention of this text to train future multicomponent distillation program authors, such detail will not be gone into; it is available in other texts, e.g. Holland, or in the original literature.

RELAXATION METHODS

A third basically different method of solving the multicomponent distillation calculation problem is by the 'Relaxation Method'. This involves locating the steady state profile of a distillation by integration of the dynamic model of the system to 'infinite' time, where 'infinite' time is taken as a time large enough to locate the asymtotically achieved profile, (see Figure 4.4).

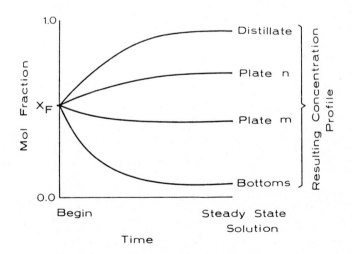

FIGURE 4.4.
THE RELAXATION METHOD: INTEGRATION TO STEADY STATE.

The distillation model is so written that differential equations describe the composition changes that will occur to the liquid hold-up on each theoretical plate. These composition changes are a result of differential equations representing component flows on to each plate and out of each plate. When these components flows are not equal, composition changes occur until a steady state condition is reached where all component flows out of the plate equal flows on to the plate, for each plate, for each component. This steady state condition is equivalent to the algebraic solution of the distillation equations. However, the solution can be reached without the difficulties of non-linear simultaneous equation solving and the other instabilities associated with the algebraic distillation calculation. The integration to a steady state is known to be a much more reliable method of locating the solution than any other available method.

This reliability is easy to appreciate. The method is effectively simulating the start-up of a column. We start with the feed composition as the liquid composition on each plate, and integrate to the equilibrium condition - comparable to the physical start up of the column.

The disadvantage of the method is that the integration of the column to a steady state takes more computing time than the algebraic location of this point (assuming that it can be located). Computer times between 10 and 100 times that of the algebraic calculation are required to locate the solution by the relaxation method.

Since this extra computing is a high price to pay for all runs, simply to ensure that an answer can be located in 100% rather than 95% of the cases, the relaxation method is not generally found as a chosen algorithm in developed distillation packages.

The detailed development of the different equations will not be demonstrated here. They are very similar to the differential equations necessary for batch distillation with modifications for the feed and boiler balance equations. The batch distillation equations are developed in the next chapter.

4.3b. CONVERGENCE DIFFICULTIES

The convergence of the distillation equations causes difficulties for two basic reasons. Firstly, the composition variables x and y are mole fraction, and are constrained to be between zero and 1.0. Outside this region the system is not only physically unfeasible, but the equations are no longer numerically soluble; this means that it is not even possible for some variables to drift out of their constrained range temporarily during the search for the convergence, because this will cause computational failure.

The second reason for convergence problem is associated with the counter-current nature of the process, and the high magnification per stage that can occur. If we take a very light component (for instance hydrogen) in the feed, this will effectively all be present in the top product, enabling a hydrogen concentration to be determined by mass balance. The hydrogen in the bottom product will be zero. Unfortunately our mathematical solution requires that this 'zero' be given an accurate value, because it is present in the equation set that must be solved. In the case of a plate-to-plate method, it is clear that an accurate bottom composition must be found so that when it is followed up the column to the top, the mole fraction determined by the feed mass balance is also achieved. Since the increase in composition per stage is very high, we are faced with the accurate estimation of trace bottom compositions of lights, where small errors produce mole fraction violations higher in the column.

For example, a bottom composition of 0.000000001 may result in a mole fraction above 1.0 higher in the column and a computational failure, whereas 0.000000095 be the correct converged result.

Hence, the plate-to-plate calculation can be expected to present difficulties when very light components are in the system. Azeotropes also cause problem because an azeotropic top product may have have a composition of 0.9000, and a good separation may achieve 1% of its value 0.8900. In iterating to converge this value, any estimate over the azeotropic composition - 0.9100, for instance - will cause a divergent situation and a very large residual. This mathematical non-continuity of the system makes convergence very difficult. This situation is made worse because the azeotropic composition is not known prior to the calculation; it is an intermediate variable resulting from the calculation, so no 'a priori' composition constraint can be set.

A further problem in convergence occurs with highly non-ideal systems. Here the activity coefficient is a function of composition, and this produces a high non-linearity of the system where fluctuations in composition during iteration are magnified by fluctuations in activity coefficient, which result in convergence problems.

It is clear that the plate-to-plate methods have notably greater instability problems than the matrix methods, but, against this, the matrix method must solve for a larger number of variables, which itself can cause convergence problems.

All methods have their problems, and despite many years of research, no efficient algorithm has yet been developed that is reliable for all types of distillation. Each has its strengths and weaknesses, and when claims are made that a superior algorithm has been developed, this usually means that the author has not yet located or tested his algorithm with those systems that indicate his algorithm's weaknesses!

Because of this situation, a good, robust industrial distillation program often contains a number of different convergence algorithms, each covering the others weaknesses, so that should one algorithm fail, the next algorithm will automatically be tried.

4.3c. COMPUTER TIME

The design of a distillation tower with a simulation program requires a large number of individual computer runs to locate suitable 'optimum' conditions; twenty runs will be a reasonable estimate.

If the distillation is part of a complex system of columns with recycles - for instance the light end section of an oil refinery - then we are talking of possibly four towers so linked together that recycles affect feed composition to such an extent that they are best designed together as part of a flowsheet simulation program, which itself will iterate to converge the recycles involved.

These examples show that the distillation calculation itself is going to be built upon and run many times before a final design is obtained.

Without trying to forecast too far into the future, it is possible to envisage the need to use a flowsheet program to optimise a set of 4 interconnected columns, each with particular specification and each containing an average of 50 plates, for a system containing 25 components.

What would the computational requirements for such a problem be?

(1) The flowsheet might have to run 200 times to locate the optimal conditions

(2) Each run might require 10 iterations to achieve conversion of the recycles

(3) Each column might require 5 iterations for the convergence to the product specification

This results in a total of:

$$200 \times 10 \times 5 \times 4 = 40,000 \quad \text{column calculations}$$

Now we see the importance of reliability and computer time. To solve such a problem is at present not feasible because the computer time for a single distillation is too long. Assuming that a 50 plate column with 25 components takes 1 minute computing, then our optimisation problem would take about 1 month's computing - and during that time it should not fail once.

We have a situation in which our problems are still bigger than can be handled by our present computer facilities. There is a need to reduce the computer time in multicomponent distillation, and for this reason the subject is still actively researched.

To complete our illustration, we can show why the distillation column takes about 1 minute, and which elemental calculations takes the major share of the computing:

(4) Each distillation simulation may take 10 iterations to converge the profile

(5) Each profile calculation requires a 50 plate calculation

(6) Each plate calculation requires 3 bubble point calculations to determine the plate temperature

(7) Each bubble point calculation requires 25 evaluations of vapour pressure and activity coefficients (1 for each component)

This results in:

$$40,000 \times 10 \times 50 \times 3 \times 25 = 15 \times 10^8$$

evaluations of vapour pressure and activity coefficient in the data bank.

"THE MAJORITY OF ALL DISTILLATION PROGRAM COMPUTER TIME IS SPENT CALCULATING PHYSICAL PROPERTY DATA"

Attempts to reduce the number of data bank calls by using old values has a detrimental effect on the iteration and reliability, because if updates are other than every iteration then discontinuities are brought into the system.

It is always preferable for the distillation program to attach the data bank just once to load the VLE parameters into its own program, from which it can use its own VLE software to save data bank attaches. This requires a duplication of software and is contrary to the principle of data banks, but computer time saving justifies this approach.

There is a great challenge and considerable potential for reduction of computer time in distillation calculations, and there will be no difficulty in finding new uses for faster distillation simulations.

4.4 MULTICOMPONENT SIMULATION - WITH ENTHALPY BALANCE

In Section 4.3 we have made the assumption that the molar flow in a column remains constant: 'constant molar overflow'. This is a simplification which has been made for instructional purposes. In reality, the flow from plate-to-plate is the result of an energy balance. The main factors relating the energy balance on a single plate to result in flows are:

(1) The liquid enthalpy change, i.e. the sensible heat change between the inlet and outlet liquid, due to changes in boiling point and composition.

(2) The vapour enthalpy change, i.e. the change in molar latent heat of the mixture due to composition change.

(3) The heat loss from the plate.

Given parameters for the calculation of enthalpy for all the components (and given correlation for enthalpy of mixtures if there are heats of mixing problems),it is possible to evaluate the enthalpies of the 4 streams associated with each plate. Given also a heat loss per plate h, we can say on every plate there must be an enthalpy balance:

$$\text{Total Enthalpy in} \; = \; \text{Total Enthalpy out}$$

$$H^V_{n-1} V_{n-1} + H^L_{n+1} L_{n+1} = H^V_n V_n + H^L_n L_n + h \qquad (4.10)$$

where H^V_{n-1} denotes molar enthalpy of vapour stream n-1 etc..
This equation now replaces our assumptions that:

$$V_{n-1} = V_n \quad \text{and} \quad L_n = L_{n+1} \qquad (4.11)$$

implied by the constant molar overflow model.

The overall mass balance of course still applies;

$$V_{n-1} + L_{n+1} = V_n + L_n \qquad (4.12)$$

We are therefore introducing two more variables per plate, V_{n-1} and L_{n+1} and two more equations per plate (4.10) and (4.12).

The solution of this extended set of plate equations is not straightforward. It is not possible to so sequence the equations per plate that the solutions can be achieved analytically, and the total set of plate equations (enthalpy, mass balances, component balances and VLE) must be solved simultaneously using a suitable iteration method.

This extends the computer time, it also creates an extra demand for data; now enthalpy data for all components must be supplied.

4.4a COMPARISON OF CMO AND ENTHALPY BALANCE RESULTS

Clearly the enthalpy balance approach causes extra computation and a demand for extra data. Is it really necessary to use it?

The most important effect of using enthalpy balance is in the calculation of vapour flow in the column. Particularly because of the difference in latent heats between the components being separated, the vapour flow can change significantly over the column. For example, in an acetone/water separation with molar latent heats being 32,200 and 40,600 kJ/kmol respectively, the flow in the bottom of the column is only 80% of that at the top of the column to transfer the same amount of heat. This ratio can be as high as 4:1 for some more exotic systems, leading to very large changes in vapour loading down the column.

For vapour loadings and column flooding calculations it is therefore important that these differences in enthalpy be accounted for.

The difference between enthalpy balance and constant molar overflow calculations does not appear to be so marked for the calculation of number of plates. This comment is based upon random observations, and, to the author's knowledge, there is no systematic work published comparing results of the two approaches. The reason for this independence would seem to be that the VLE data is hardly altered by the enthalpy balance, which introduces only second order corrections to the liquid composition to maintain both the mass and enthalpy balance.

4.4b A MODIFICATION OF CMO RESULTS TO PREDICT VAPOUR LOADING

Assuming that a CMO calculation gives us an acceptably accurate estimate of the number of plates required, we need to get an improved estimate of the column vapour loadings before the column can be sized. This can be done by making an enthalpy balance across the whole column. Figure 4.5 shows the flows leaving the top and bottom of the column and the heat losses from the tower h.

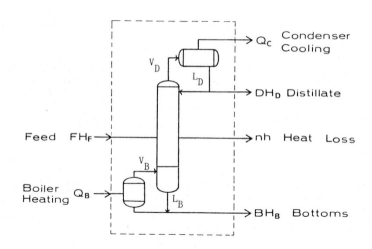

FIGURE 4.5
ENTHALPY BALANCE OVER A DISTILLATION TOWER.

Given a feed rate F, and distillation rate D, from a mass balance over the column we can say what the flows around the columns are;

$$L_D = RD \qquad (4.13)$$

$$V_D = (R + 1)\, D \qquad (4.14)$$

$$L_B = V_B + F - D \qquad (4.15)$$

V_B we do not know, since this must come from the enthalpy balance.

Taking enthalpy balance around the column itself:

Total Enthalpy in = Total Enthalpy out

$$H_F F + H_D^L L_D + H_B^V V_B = H_B^L L_B + H_D^V V_D + \Sigma h \qquad (4.16)$$

where H corresponds to the enthalpy per mole of mixture at the point in the column and temperature denoted by the sub- and superscripts.

Substituting Equations (4.13) to (4.16), we get:

$$H_B^V V_B = H_B^L (V_B + F - D) + H_D^V (R+1)D + \Sigma h - H_F F - H_D^L RD \qquad (4.17)$$

and

$$V_B = \frac{H_B^L (F-D) + H_D^V (R+1)D - H_F F - H_D^L RD - \Sigma h}{(H_B^V - H_B^L)} \qquad (4.18)$$

Equation (4.18) therefore enables the vapour loading at the bottom of the column to be calculated for a column with a specific top product rate D and reflux rate R.

The column diameter and internals design can then be calculated using both V_D and V_B to determine whether a single diameter or internals design can be used over the whole column. Should there be large differences in vapour flow between the top and bottom of the column it may be necessary to calculate flows at other intermediate points in the column to determine where maximum and minimum loadings occur before a design can comfortably be made.

4.5 UNDERSTANDING THE PROFILE

Now that the multicomponent system has many more variables than the binary system, the concentration profile down the column is much more important because it delivers so much information concerning the separation conditions. Once a proper understanding of the profile is reached, it is so much easier to decide on the next condition for an improved design.

Figure 4.6 is a profile for a quaternary system, operating under good conditions, with 2 clear 'key components', i.e. important with respect to specification and also as major constituents in the feed. The other two components are minor in amount, one heavier than the two keys and the other lighter.

Notice that the profiles of the two keys are similar to that of a binary profile. The same observations apply:

- long section with no changes in concentrations on both sides of the feed denotes too low a reflux ratio

- no change of key components at one side of the feed denotes an error in the feed position

- too sharp a profile denotes an uneconomically high reflux ratio

- inadequate separation when other factors are in order denotes inadequate number of plates. The profile shows where the extra plates should be put.

- incorrect mass balance (incorrect specification of distillate rate) shows itself with "dead" column sections in half of the column.

The light and heavy components do not interfere with our analysis. At low
concentrations they show points of minor interest (their profiles are not ideal, they will have long sections where no concentration changes occur), the
feed will be entirely wrong, and there will be no plate that is equivalent to
the feed composition because of the extra degrees of freedom provided by 3rd
and 4th components.

As these non-key components increase in concentration in the feed, the same
arguments apply, except that the concept of a key becomes fuzzy when it is no
longer a major constituent in the top or bottom product.

The major change occurs when the keys are not "adjacent", i.e. a minor component lies between the two keys in volatility.

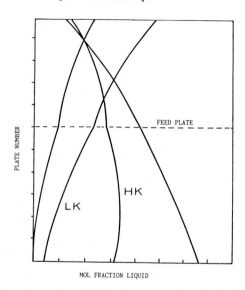

FIGURE 4.6
QUATERARY SEPARATION WITH ADJACENT KEY COMPONENTS

Figure 4.7 shows a profile for such a system containing three components with
the middle volatility being a low concentration component whose final destination is not of importance. Now the profile is much more complicated. The non-
key component is being held back from being in the distillate because it is
heavier than the distillate, and it is held out of the bottom because it is a
light with respect to the bottom product. It has to build up in concentration
within the column until, at the high concentration levels, the quantity that
can escape with the products is in mass balance with the quantity being fed.

Hence, concentrations of this component rise in the column. There are 2
clear sections marked as A-B, and B-C on Figure 4.7. In section A-B, the major
separation is between the heavy and the non-key. The light key is not involved
in the separation and remains at a near constant composition. Part B-C is a
separation between the light key and the non-key, and the light key and the
heavy key.

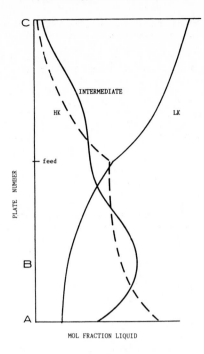

FIGURE 4.7
TERNARY SEPARATION WITH NON-ADJACENT KEY COMPONENTS

This situation is more or less clear depending on the relative volatilities of
the 3 components, their concentrations and the split required of the non-keys
between top and bottom products. When the destination of the non-key is of no
consequence, conditions can be found where this build-up of non-key in the
column is minimized, but as soon as a specification is also set on the split of
the non-key, then, if this is not in harmony with the relative volatility,
this can lead to amazingly high build-ups of the non-key in the column. For
instance, if the non-key is rather closer to the light key than the heavy key
in volatility, but nevertheless the majority is required to be in the heavy
product, then very high build-ups occur. An example in the author's experience
is of 40% concentration in the column, with only a 0.2% in the original feed.

These build-ups, and column sections that are not associated with the separation of the key components, lead to increased numbers of plates. Two or more
times the number of plates may be required, compared with a calculation
neglecting the presence of the non-key.

Of course when we are talking of high build-ups in concentration in the column,
and specifications also being put on the non-keys, and this specification sometimes being the major separation, then we should really ask if the non-key is
really a key. The concept of "key components" begins to break down, and although it is very important in many instances, it really cannot be applied to
such situations as that now being described, and we must accept that we have a
distillation problem with 3 important components.

On top of these inversions in the profile, there are the other profile characteristics described in Chapter 3. Large sections with virtually no concentration change denotes too low reflux ratio (note that in many instances a "key"
may remain constant over a large section while concentration changes with the

other two components, even at optimum operating conditions). Incorrectly specified distillation rate gives mass balance problems and sections of "dead" column where no change occurs.

In principle, the same arguments as described in Chapter 3 hold: that the feed
point is so chosen that the rate of separation of the keys above and below the
feed are the same (and therefore combine to a maximum). However, now we are
not certain which really are the keys and thus we are never sure which curve to
use for the feed point location. In effect, a compromise is required for all 3
curves, and the feed should be moved to favour the particular components separation that is causing difficulty in meeting specifications. Changing the feed
point changes the split of the middle component between top and bottom products.

Particularly important in these 3 component systems is the mass balance problem. If too much mid-component appears in the tops, it may be that more separation is required above the feed, but it is equally possible that the need is
for more separation below the feed, so higher boiler concentrations can be
achieved and hence, by mass balance, low top concentrations can be obtained.

Only by inspecting the profile can the correct actions be taken to improve the
design.

4.6 LOCATING THE FEEDS AND SIDESTREAMS

Feeds

As can be judged from the previous discussion, there is no rigorous rule for
locating the optimum feed point position for a multicomponent separation comparable to the simple binary rule that the feed enters at the plate with nearest
composition to the feed. A guideline is that the plate should have the same ratio of concentration of keys as the feed. The presence of the other components
in the system means that it is highly unlikely that a plate can be found for
which the ratio of all components to each other is the same as the feed, and we
are immediately in difficulties when we decide that there are more than 2 important components in the column. When this is the case, a good compromise is
called for.

Having made a preliminary estimate, a simulation run will produce a profile
that can be inspected and revised conditions, including revised feed point location, can be tried. Generally the feed point position has a second order effect on the separation. Except for grossly misplaced feed points, changes are
small and shifting the feed point 10 or 20% up or down the column generally
produces only minor product concentration changes. "Optimal feed point
location" is done after locating the correct distillate rate, reflux ratio and
total number of plates.

Complex distillation column often have more then one feed. The rule for each is
to make an initial estimate based on ratio of important components, and then
investigate changes by simulation. What could become a very difficult
multivariable optimization problem is simplified in that the product
compositions are relatively insensitive to changes in feed point location, and
the profile gives extra information with which to make decisions.

Sidestreams

Product withdrawal through sidestreams is possible in multicomponent system,
because, as the profile shows, considerable variation of liquid composition occurs within the column. When pure products are not required, but only liquid
mixtures with particular properties, withdrawal of a sidestream can produce other useful products in addition to the top and bottom products.

The sidestream should be located at the position where the composition in the column best represents the required properties of the sidestream. The sidestream rate must also be specified.

The very removal of the sidestream alters the mass balance, liquid/vapour flow ratios and hence column profile. In particular, that best point in the column will be less good: the concentrations of the required component will be lower because it is being removed and so does not have the chance to build up to the high level that may have looked so promising when the original profile without a sidestream was investigated.

Hence the removal of the sidestream alters the condition to such an extent that further simulations have to be done and new take-off rates, reflux ratios and number of plates determined.

Sidestream take-off is common practice in the petroleum industry; it is used in the chemical industry to give product improvement as described in Chapter 6.

4.7 ROBUST, FLEXIBLE DESIGNS

As with the binary design, no design is complete until the alternative futures have been examined, and any reasonably likely future studied with the proposed design.

Since multicomponent distillation calculations are simulations, this facilitates the study in that the proposed equipment is specified and the new feed fed to it. The result of the calculation is the performance of the proposed design with the new feed.

If the result is unacceptable, minimal changes should be proposed that enable possible future conditions to be handled on the proposed design. This is a new design problem, with the constraint that the changes to the first design should involve minimal capital cost.

4.8 AN EXAMPLE

A column is required to produce chloroform (CF) of adequate quality for Freon 22 and PTFE manufacture. It must therefore meet a specification of 0.1000 wt% CCl_4 (CTC) and 0.005 wt% CH_2ClBr (CBM). 90% recovery of chloroform as product is required.

The column feed is 70 kmol/h liquid with a temperature of 60° C and concentrations of 0.20 mf CCl_4 and 0.002 mf CH_2ClBr.

Data:

Antoine Coefficient	A	B	C	mol wt
Chloroform	12.4932	4833.91	52.7	118.5
Chlorobromomethane	12.4682	4857.08	48.8	129.5
Carbon tetrachloride	12.2540	4866.59	47.9	154.0

For the sake of this example, the system can be considered to be ideal.

We must start with a proposed design with which to carry out our first simulation runs. Since the CBM has a vapour pressure that lies between the two major components (see Table 4.1), we cannot expect Fenske to give an accurate result, and we will also have difficulty in defining which are the keys.

The way out of this dilemma is to make a few Fenske runs, using different pairs of keys, and then select a column design to simulate having inspected all the results.

Before we can run the program, there is need to do a little manual mass balance work, firstly to arrange the input data for the design programs, but also to get a feel for the problem before we use the computer.

Firstly, converting wt% to mole fraction we get:

$$0.10 \text{ wt\% CTC in CF} = 0.00077 \text{ mol fraction}$$

$$0.05 \text{ wt\% CBM in CF} = 0.00046 \text{ mol fraction}$$

We are given the feed compositions and flow, and distillate compositions and flow (90% recovery of CF).

By material balance we can calculate the bottom flows of each component, and hence the bottom composition. The results of this manual calculation are summarized in Table 4.1.

TABLE 4.1
PREPARATORY CALCULATIONS FOR THE EXAMPLE

	V.P. at 80 Deg C (bar)	Feed mol fr.	kmol/h	Distillate mol fr.	kmol/h	Bottoms mol fr.	kmol/h
CF	1.79	0.798	55.86	0.9987	50.27	0.283	5.58
CBM	1.46	0.002	0.14	0.00046	0.023	0.0060	0.117
CTC	1.22	0.200	14.00	0.00077	0.0384	0.710	13.96
Totals	–	1.000	70.00	1.0000	50.33	1.000	19.66

The next step is to run the 3 separate Fenske runs taking the data from Table 4.1. Note that the program can only be run with adjacent feeds, thus for the case with CF/CTC as the two keys, the system must be considered as a binary.

Table 4.2 gives the results of the 3 Fenske runs, using the program FENGIL (see Appendix 4).

It is reasonable to take the most pessimistic of these cases as the first trial for the full simulation. Since the most tight specification is the CBM content in the distillate, we suggest more plates above the feed than below;

First trial; 41 plates, reflux ratio 4.2, feed plate 6.

This case was then simulated with the program DISTIL (see Appendix 4). The result of this simulation showed very little separation of the CBM. The reason for this could easily be seen from the CBM profile (see Figure 4.8). The feed

point is so low that there is no opportunity for the CBM to build up concentration in the bottoms - and so it must come out of the top!

If we try increasing the feed to plate 20, a much more satisfactory result is
obtained, as the second curve in Figure 4.8 shows.

TABLE 4.2

RESULTS OF FENGIL RUN FOR THE EXAMPLE

KEYS	NO. OF PLATES	REFLUX RATIO (1.2 x MIN)
CF/CTC	17	2.0
CF/CBM	41	4.2
CBM/CTC	16	2.0

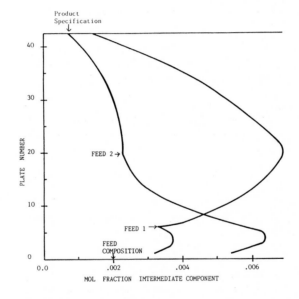

FIGURE 4.8
CBM PROFILES FOR THE EXAMPLE FOR 2 FEED LOCATIONS

This case shows that we have a better product than specification, in terms of
both CTC and CBM, and the profiles show no problem of shortage of reflux. Hence
to get closer to specification, the reflux ratio was reduced to 3.6 and still
the product was found to be on specification.

We now have the feed point correctly located (to a first approximation), we
have achieved specification, and the reflux as shown by the profile is not excessive. The distillate take-off-rate in the example causes no problem because
it concerns only a 90% recovery of light product. This becomes more of a prob-

lem when 99% lights recovery or tight specification on the light contents of the bottom products are required.

Our next step will be to do a little optimization to see whether the reflux ratio/number of plates relationship is optimal, similarly to the example in Chapter 3. Then, at these new conditions we should reconsider the feed position. Firstly by looking at the profile, then, if necessary, by a case study, shifting the feed by a few plates to observe the result.

Following this, we consider flexibility, control and alternative futures, in a parallel fashion to Chapter 3.

The whole study, properly done for a column that is going to be built, takes a number of weeks. Since this is only a demonstration, we will not go through these stages.

4.9 FURTHER READING

Fredenslund A., Gmehlin J. and Rasmussen P., 1977. Vapour-Liquid Equilibrium
                using UNIFAC, Elsevier.

Henley E. J. and Seader J.D., 1981. Equilibium-Stage Separations Operations
                in Chemical Engineering, Wiley.
                - good on multicomponent short-cut methods.

Holland C.D., 1981. Fundamentals of Multicomponent Distillation, McGraw-Hill.
                - detailed discussion of some convergence methods.

King C.J., 1980. Separation processes, McGraw-Hill.
                - devotes attention to short-cut methods.

Naphtali L.M. and Sandholm D.P., 1971. AIChEJ, 17, 148.
                - describes the convergence method used in the DISTIL program.

Perry R.H. and Chiltern C.H., 1973. Chemical Engineers' Handbook, McGraw-Hill.
                - particularly for gas-liquid contactors.

4.10 QUESTIONS to CHAPTER 4

(1) Why are rigorous multicomponent distillation calculations forced into the simulation mode?

(2) Why are short-cut multicomponent methods so useful?

(3) Why are there 3 basically different methods for rigorous calculation of multicomponent distillation?

(4) What are the particular problems with each of these methods?

(5) In what program section is the major computational load in a multicomponent distillation calculation?

(6) What are the basic differences between the CMO and the Enthalpy Balance approach?

(7) What information is contained in the profile? Why is its understanding so important for multicomponent distillation?

(8) How are the position of the feeds and the sidestreams located?

Chapter 5

BATCH DISTILLATION

5.1 THE USES OF BATCH DISTILLATION

Batch distillation is much simpler in concept than continuous distillation. A mixture is charged into a boiler, and this mixture is heated to vaporise it progressively. The vapour is condensed and collected in a series of fractions. Because the vapour has a higher concentration of the light component than the corresponding liquid, the condensed fraction has initially higher light concentrations than the original mixture, and the later fractions (or the part left in the boiler) have lower concentrations. This single vaporisation stage is called a 'simple' or 'Raleigh' distillation.

Because the separation from a single equilibrium stage is generally not sufficient, it is usual to incorporate a number of stages by inserting a column between the boiler and condenser, filled with mass transfer internals down which a part of the condensate flows as reflux. This has then exactly the same function as one half of a continuous distillation column, and a multistage equilibrium separation can be achieved, the degree of which can be varied by the reflux ratio of the returned condensate and by the quantity of mass transfer packing or trays in the column.

The operation of such a column is still very simple: a mixture is charged into the boiler, vaporised through the column, part of the condensate returned as liquid, and the remainder collected as a series of fractions of differing composition.

The system may be simple to operate, but it is traditionally abhorrent to a classical chemical engineer; conditions are not constant, the batches must be replenished and the series of fractions monitored, switched and discharged; manpower costs are high and equipment volumes large compared to achieved throughput – furthermore it is difficult to calculate! This leads to a general rule that, whenever possible distillations (and processes) shall be made continuous.

Such views have been propagated by oil refining-specialists, born out of a lack of understanding of other industries. In many industries, batch processes are the preferred type of process, based on economics and operational criteria. Batch processes have advantages over continuous distillation and each should be judged on its own merits.

The outstanding advantage of batch distillation is that one column can separate a mixture containing any number of components into its pure components. To achieve a similar result with a continuous distillation system requires n-1 columns for a system of n components.

The second advantage of batch distillation is that it is very much more flexible than a continuous column, and more suitable as general purpose equipment for purification and solvent recovery within a plant where numerous distillation tasks occur. Any speciality chemical plant, producing batchwise 50 or 100 different products in campaigns will have numerous distillation tasks, which can all be achieved by a single general purpose distillation plant (with an adequate number of stages) within the works.

For small tonnages, it is no disadvantage if the equipment volumes per unit are larger for batch than for continuous distillation. A small plant may require a 20-mm-diameter continuous column or a 60-mm batch column, but at these sizes the saving in capital is negligible in comparison with costs such as the controls, storage and buildings. For small tonnages, the disadvantage of the larger equipment size of batch distillation is not relevant. Naturally, as plant capacities increase, there comes a point where the batch equipment become expensive because of its size, and the additional manual work is also relevant. That is the time to use a continuous distillation system.

The choice of batch or continuous distillation should be made simply on economics and plant operability grounds, determined case by case. There is no good rule except to say that it would be very surprising to see a batch distillation plant with a capacity of more than 10,000 tons a year, or a continuous plant handling less than 10 tons per year.

5.2   THE TOTAL BATCH SYSTEM

Continuous distillations operate at steady conditions. A column size can be determined to achieve the required output, and then the boiler and condenser are sized appropriately.

Batch distillation conditions are not constant. Reflux ratio, compositions, and temperatures change throughout each cycle. The equipment size is required to achieve an suitable cycle time for a given batch size; the batch size is usually given by the rest of the batch process and a suitable batch time will be related to the batch time of the rest of the process. It is not possible to determine this cycle time without considering the conditions during the whole cycle. This requires the control policy to be specified: will operation be at a constant reflux, or an increasing reflux? Furthermore, is it policy to have the column always limiting and oversize the condenser and boiler? Or should an optimal sized condenser and boiler also be installed?

In an optimally-sized system, the condenser is limited at the beginning of the cycle when the light components have to be condensed, the column diameter at mid-cycle, and the boiler heat transfer at the end of the cycle where boiler temperatures are highest. In such a system, the cycle time is dependent on the condenser and boiler area as well as the column diameter and reflux policy. This simple step-wise design approach of the continuous distillation cannot be applied, and a good batch distillation design is concerned with a design of the batch distillation systems as a whole. Figure 5.1 shows a total batch distillation system.

This, of course, does not make for easy design, and a good design of a batch system must be a system study of the whole equipment. System studies are best achieved by means of simulation. A system is defined and the performance of this system is calculated, the resulting performance is inspected, a modified system is proposed, its performance recalculated and the cycle repeated until a suitable total system design is achieved.

The centre of the system is the column itself. It is defined by its vapour capacity (through its diameter and internals characteristics) and number of theoretical separation stages (through the performance of the column internals). Following this, reflux control must be specified. Will a constant reflux ratio be used and the fraction changed when a cumulative fraction composition just meets the defined product specification, or will a constant top composition (or temperature) be held by increasing reflux ratio until an uneconomically high reflux ratio is attained?

Furthermore, the vapour rate during the distillation must be determined. What limitation to the achievable maximum column vapour rate is introduced by the condenser area at the beginning of the cycle, and by the boiler area at the end

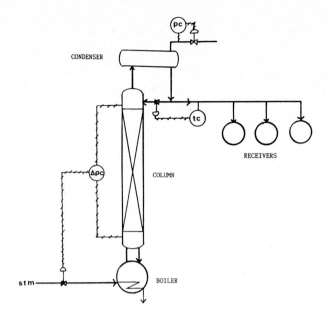

FIGURE 5.1.
THE TOTAL BATCH DISTILLATION SYSTEM

of the cycle? The maximum capacities of these two items depend upon the overall heat transfer coefficient and temperatures during use. The fraction of the time each item is limiting is determined by the column top and bottom temperatures.

When all these points have been defined, the appropriate simulation calculation can be made and the batch cycle time determined.

5.3   THE BATCH EQUATIONS

The simple distillation using a single distillation stage is indicated by Figure 5.2.

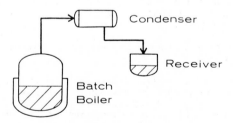

FIGURE 5.2.
SIMPLE BATCH DISTILLATION

At a given instant in time, let there be B moles of binary mixture, mole fraction x ,in the boiler. Let a small quantity be vaporised, of composition y.

Then, by mass balance in the boiler:

$$- ydV = d(Bx) \tag{5.1}$$

expanding $\quad - ydV = xdB + Bdx \tag{5.2}$

since $\quad - dV = dB \tag{5.3}$

$$ydB = xdB + Bdx \tag{5.4}$$

re-arranging and integrating:

$$\ell n \left(\frac{B_o}{B_f} \right) = \int_{x_f}^{x_o} \frac{dx}{y - x} \tag{5.5}$$

where subscript zero denotes the initial condition and f the final.

This is the basic equation, which represents the simple, single plate distillation.

Given a relationship between y and x, the RHS of Equation (5.5) can be integrated to enable the time profile of x, y and B to be evaluated.

In general, this integration can be carried out graphically, working with the x/y curve if no other data is available. If a VLE model is available, there is a mathematical relation between x and y that can sometimes be integrated to give an analytical solution to Equation (5.5). For example, if we assume ideality, with a constant relative volatility, then the relation between x and y is defined by Equation (2.8). This results in an analytical solution of Equation (5.5):

$$\ell n \left(\frac{B_f}{B_o} \right) = \frac{1}{\alpha - 1} \ell n \left(\frac{x_f (1 - x_o)}{x_o (1 - x_f)} \right) + \ell n \left(\frac{1 - x_o}{1 - x_f} \right) \tag{5.6}$$

This treatment indicates the basic constituents of any batch distillation calculation; the description of the VLE, the mass balance equation defining the boiler and the product fractions, and the integration of these equations over the course of the distillation.

The above theory was extended by Smoker and Rose to the batch fractionation case, where a column with a number of theoretical plates is placed between the column and the condenser, and some condensate is returned as reflux.

The same argument leads to the development of the equation analogous to Equation (5.5):

$$\ell n \left(\frac{B_o}{B_f} \right) = \int_{x_{B_f}}^{x_{B_o}} \frac{dx_B}{y_D - x_B} \tag{5.7}$$

which relates the bottom composition quantity to the composition of the distillate removed from the system.

The relationship between y_D and x_B is now defined by a fractionation calculation for a given number of plates and a given reflux ratio, with a given VLE model. Relations between x_B and y_D can be obtained by and using the McCabe-Thiele diagram, at a constant L/V, reading off pairs of values of x_B and y_D for various stages of the distillation. This data can then be used to construct a graphical solution to Equation (5.7).

Given an initial boiler charge of B_0 kmols, being distilled with a reflux R and boil-up rate V kmol/h, which is distilled until the boiler composition is x_{B_f}, what is the mean distillate composition and how long does the distillation take?

Using the McCabe-Thiele method, the integration of the Equation (5.7) can be done graphically and hence B_f can be determined.

The mean distillate composition is given by mass balance:

$$\bar{y}_D (B_0 - B_f) = B_0 x_{B_0} - B_f x_{B_f} \qquad (5.8)$$

The time required to collect (B_0 - B_f) kmols distillate is given by the molar flows around the column:

$$V = (R + 1) D \qquad (5.9)$$

$$D = \frac{V}{R + 1} \qquad (5.10)$$

hence the batch time is:

$$\frac{(B_0 - B_f)(R + 1)}{V} \qquad (5.11)$$

The Smoker Method enables an estimate to be made of the distillation time for a binary system under fixed operating conditions. The results of this analysis may be adequate for very small equipment where oversized equipment is of no economic consequence, but, in cases where an investment is to be made, a more thorough analysis is called for: not only to ensure that equipment costs are minimized, but also to confirm that the proposed equipment will meet the the specified cycle time demanded of it. Even though the distillation equipment may be of low cost, the economic loss through a restricted output due to inadequate distillation capacity may be very considerable and may be the real justification for a more accurate distillation design study than the Smoker method can provide.

Any more thorough treatment of batch distillation than the Smoker method must be based on numerical simulations with the computer. As always, this extends the horizon remarkably, enabling non-ideal systems and distinct operating policies to be included in the study. If necessary, the study can be extended to the simulation of the total distillation system itself, and so we are meeting our earlier defined objective of determining the total cycle time for a total system.

The numerical simulation consists of connecting a batch boiler to the rectification section of a distillation column and integrating over time until the light component has been adequately removed from the boiler. Figure 5.3 indicates the basic procedure.

The determination of the composition profile in the column is exactly the same as the multicomponent distillation profile for a continuous column, and the solution methods available are the same. For simple systems (e.g. binary) a plate-to-plate is adequate. Alternatively, various matrix methods can be used, as can the relaxation method, which integrates the profile to a steady state.

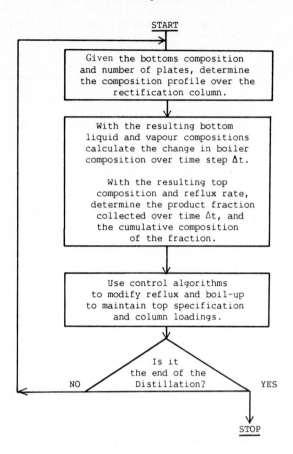

FIGURE 5.3
FLOW CHART FOR BATCH DISTILLATION CALCULATION PROCEDURE

It is now easy to appreciate the problem in batch distillation design. The calculation is a degree more difficult than continuous distillation, because the
distillation model has to be integrated over the whole distillation cycle. One
hundred to one thousand continuous distillation design calculations are necessary to make correctly a single batch calculation. Add to this the fact
that the batch distillation is concerned with small tonnages and small equipment, and the problem is clear:

"THE BATCH DISTILLATION DESIGN PROBLEM LIES IN CHOOSING THE APPROPRIATE DEGREE
OF DETAIL TO ENABLE A REASONABLE DESIGN TO BE ACHIEVED FOR DESIGN COSTS WHICH
ARE COMMENSURATE WITH THE EQUIPMENT COSTS INVOLVED"

5.4 LIQUID HOLD-UP

A major difference between batch and continuous distillation concerns the effect of liquid hold-up on the system, both in the column itself and around the
condenser system.

Because the continuous system is in steady state, the hold-up has no effect on the analysis. Any quantity of liquid held up in the system has no effect because the concentrations remain constant and thus the hold-up concentrations do not affect the mass flows in the system. In batch distillation the compositions are changing with time, and the liquid hold-ups in the system are sinks of material which modify the rate of change of the compositions in the system.

A second effect of liquid hold-up in batch systems is on the material balance equations. Material held up in the system is not in the boiler. Unfortunately, liquid held up in the system does not have the same composition as the liquid in the boiler, but it has consistently high concentrations of the lighter components. Hence, the boiler, by mass balance, has a lower concentration of light than would be expected from an analysis neglecting hold-up. This means that the separation is made more difficult because of liquid hold-up, and towards the end of the distillation fraction this effect is of major importance.

Furthermore, this material held up in the column at the end of the product fraction has to be collected in the intermediate fraction for recycling. Hence, liquid hold-up increases the intermediate fractions and reduces the yield of pure products per cycle.

5.4a THE EFFECT OF LIQUID HOLD-UP

Of the three effects of liquid hold-up, it has once been reported that the delaying effect of the hold-up in the profile has a positive effect on the performance in the column, i.e. the separation is easier than would be predicted by an analysis excluding hold-up. This conclusion is by no means obvious; it is not generally observed, and it would be foolhardy to make the assumption that this positive effect of hold-up will negate the other detrimental effects, unless it has been justified by extensive simulation.

The remaining two effects are detrimental, and the overall impression of batch distillation practitioners is that hold-up is bad. How bad depends entirely on what the system is, what performance is required from it, and how much hold-up is involved.

A packed column, with little hold-up per theoretical plate, easy separation for a 2 or 3 component mixture, and all component being present in similar amounts, would most probably perform exactly as predicted by a calculation ignoring hold up. However for a tray column handling a difficult system, with a badly designed condenser system, separating a mixture containing some components in low concentration, then the column may not produce a single drop of on-specification product of the minor components because of hold-up.

It is most important that the effects of hold-up be predicted in all but the most elementary design studies. The following example shows the effect of hold-up. The method used for this calculation is explained later in this chapter, but it is included at this point to demonstrate the effect of hold up.

EXAMPLE

Consider the separation of a mixture of methanol/water/DMF in a 6 stage tray column. The total cycle was simulated, with and without calculating the effect of hold-up. The hold-up, which in this case was considerable because it was a tray column, effects the reflux ratio policy and the cut-off point of each fraction, giving lower yields and larger intermediate fractions requiring recycling. The overall consequence is a lower yield of product per cycle (or an extended cycle time if the intermediate recycle is included) and a considerably higher energy requirement per batch: firstly, because the increased recycle of intermediate fraction, and secondly, because all reflux ratios had to be higher because of the depleted lights concentration in the boiler.

The results of the study are summarized in Table 5.1.

TABLE 5.1
EFFECT OF HOLD-UP IN BATCH DISTILLATION

	plate column (with hold-up)	packed column (no hold-up)
1st product fraction (95%)	45.6 kmol	47.1 kmol
1st intermed. fraction	7.8 kmol	5.6 kmol
2nd product fraction (95%)	22.1 kmol	23.7 kmol
2nd intermed. fraction	0.8 kmol	0.2 kmol
Bottom product (95%)	19.6 kmol	19.4 kmol
Total time	8.3 h	7.7 h
Total energy	7.9×10^6 kJ	6.7×10^6 kJ

5.4b PREDICTION OF LIQUID HOLD-UP

For batch distillation therefore, it is important to know the liquid hold-up in an operating column. For plate internals, the liquid is held up in the froth on the trays and in the liquid in the downcomers. Chapter 7 shows how the froth height and void fraction can be predicted, as well as the liquid in the downcomers, so that for these internals hold-up can be predicted adequately. A normal plate design might have a hold-up equal to 15-20% of the column volume, and a plate designed for low hold-up may reduce this value to about 10%.

Hold-up in packing is more difficult to predict. It is calculable from wetted surface area and liquid film thickness, but the film thickness is dependent on many factors; liquid rate, vapour rate, liquid viscosity, etc. There appears to be very little systematic work reported correlating these factors, and so only rough estimates of hold-up in packing can be made; accurate estimates require experimentation.

As a rough guide, packing with a specific area (a) of 200 m has a hold-up of 5-10% of the column volume when fully loaded, reducing to 3-6% at lower rates.

5.4c MODELS DESCRIBING LIQUID HOLD-UP

Dynamic Modelling of the System

There is no problem in writing down equations that describe the liquid hold-up in a distillation system and incorporating these in a separation model of a complete tower.

Figure 5.4 shows a single plate with a liquid hold-up H and composition x, with the quantity of hold-up being independent of the liquid flow rate.

It is generally assumed that the liquid on the plate plate is perfectly mixed and that the vapour leaving the plate is in equilibrium with the liquid on the plate (or a fractional approach to it). For the sake of simplicity we can also assume constant molal overflow, although it is, in principle, straightforward to incorporate heat balances on the plates for the enthalpy balance approach.

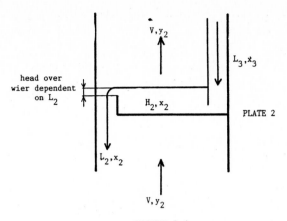

FIGURE 5.4
FLOWS OVER A SINGLE PLATE

Taking a dynamic mass balance over the plate:

$$\text{Total flow in} - \text{Total flow out} = \text{accumulation}$$

for each component we get:

$$\frac{\partial(L_3 x_{i3})}{\partial t} + \frac{\partial(V_1 y_{i1})}{\partial t} - \frac{\partial(L_2 x_{i2})}{\partial t} - \frac{\partial(V_2 y_{i2})}{\partial t} = \frac{\partial(H_2 x_{i2})}{\partial t} \qquad (5.12)$$

By describing every plate by this equation, and starting the calculation by assuming that all plates are filled with feed material, the column calculation, proceeding from bottom to top of the column, will determine the changes
that occur over a small time step δt. A numerical integration at total reflux
will then converge the column onto a composition profile, from which point a
product take-off can be made and the numerical integration will indicate the
effect of the hold-up as the material is removed and compositions change.

This method is in principle an extension of the relaxation method of converging
a steady state column, as described in Chapter 4.

The extension is to include the integration of the mass balance over the boiler
(see Figure 5.5) and a description of the accumulation of top product (see Figure 5.6).

$$\frac{\partial(B x_{io})}{\partial t} = \frac{\partial(V_o y_{io})}{\partial t} - \frac{\partial(L_1 x_{i1})}{\partial t} \qquad (5.13)$$

$$\frac{\partial(D x_{iD})}{\partial t} = \frac{V_n y_{in}}{(1 + R)} \qquad (5.14)$$

The attraction of this method is, firstly that it is a straightforward solution
of the equations describing in the system, and, secondly, that it combines the
convergence of the profile with the solution of the batch integration resulting
in a very close approach to the physical description of the processes
occurring.

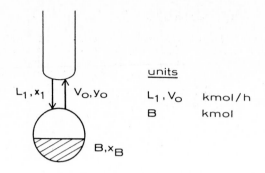

FIGURE 5.5
FLOWS AROUND THE BATCH BOILER

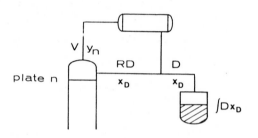

FIGURE 5.6
FLOWS AROUND THE BATCH COLUMN HEAD

The problems arise in its practical application. As already pointed out in Chapter 4, the relaxation method is not a very quick way of achieving convergence in a column, and this problem is magnified when it is coupled with the boiler mass balance equations. The resulting system is a system of differential equations with two regions for the time constants. The hold-up equations have time constants in seconds, and the boiler contents in hours. It is well known that such equations are difficult to integrate numerically ('they are stiff') and the step length must be very small and computer times correspondingly large.

The situation is intensified when packing is used for the column internals, because here the hold-up is a thin film on the packing, and the boiler may contain 10 tons of liquid! The less important the hold-up is, the more difficult becomes the calculation to predict its effect because the system of differential equations becomes stiffer!

Unfortunately, this approach does not help to solve our batch design problem of choosing a degree of detail commensurate with the equipment costs. Our systems of equations can be solved, but computer time hundreds of times longer than the time for steady state distillations are obtained and design costs are inappropriate to the size of equipment being designed.

PSEUDO-STEADY STATE ASSUMPTIONS

An established technique for treating systems with two ranges of times
constants is to assume that the set with the small time constant so quickly approaches equilibrium that during a large time step appropriate for the large
time constant system, the variables associated with the smaller time constant
remain constant.

i.e. during each large time step:

$$\frac{\partial(Hx_{ij})}{\partial t} = 0 \qquad\qquad (5.15)$$

the equilibrium values of x_{ij} are given by the plate composition profile.

Hence Equation(5.12) has a zero RHS, and the column calculation reverts to the
same set of equations as for continuous distillation.

The hold-up however does have an influence on the mass balance over the boiler.
The boiler contents are now the original contents minus the integrated
distillate removed minus all the material held up in the system!

Boiler mass balance by components at time t therefore gives:

$$(x_{iB}B)_t = x_{oi}B_o - \int_0^t Dx_{iD} - \sum_{j=1}^{N+1} (H_j x_{ij}) \qquad\qquad (5.16)$$

for i components and n plates,
H_{N+1} is the condenser hold up
and subscript o represents values at time zero.

This technique can reduce computer time from 1/10 to 1/100 of the time
for the full dynamic model. It fully simulates the negative effect of hold-up
in reducing boiler compositions and in increasing the intermediate fraction because of the material in the column. It does not simulate the reported positive
effect due to the system dynamics and can therefore be expected to produce an
overdesign rather than underdesign.

However, this approach loses information. No dynamic information is available
on start-up or controllability; such is the price to be paid for the reduced
model.

5.5 BATCH PLANT OPERATION

5.5a THE EQUIPMENT

Figure 5.7 is a flowsheet of a reasonably well instrumented batch distillation
system. A large capacity boiler fitted with either an internal coil or a jacket
for steam heating is connected to the column. Following the condenser, an arrangement of pipework and valves enables the product fractions to be collected
in separate receivers; usually collecting product and intermediate fraction
alternately.

Reflux ratio is controlled either manually or automatically from a temperature
point in the column. Direct product on-line analysis, though technically possible, is rarely economically justified.

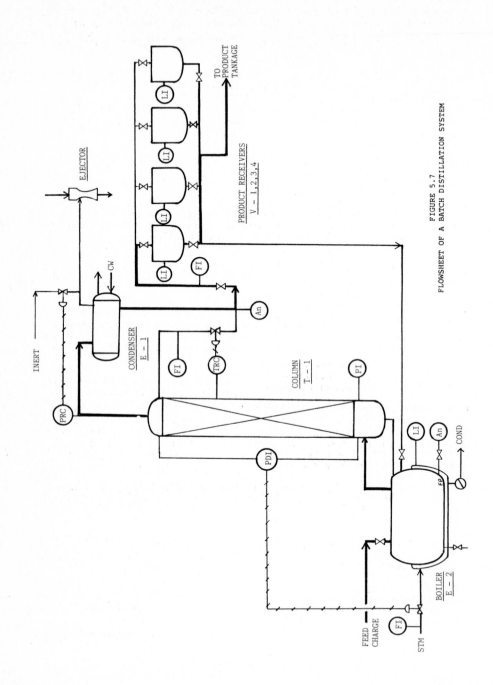

FIGURE 5.7

FLOWSHEET OF A BATCH DISTILLATION SYSTEM

In order to maintain maximum throughput it is necessary to increase the steam temperature in the boiler continually to compensate for the increasing boiling point in the boiler and the reducing heat transfer area if the heating is applied through the jacket. A usual means of column loading control to maintain the maximum throughput is by means of column pressure drop, which may be automatically or manually controlled at its maximum value by the boiler steam flow.

If the situation is reached where the heat transfer into the boiler is limiting, the capacity of the system can be improved by reducing the total system pressure. The boiling point of the boiler content drops, more heat is transferred, and the rate is thereby increased. Simultaneously, the column capacity is reduced, and so there is an optimal pressure profile to achieve once the boiler steam is at maximum pressure. Figure 5.7 shows a proposed control scheme to achieve this.

5.5b THE REFLUX POLICY

For a single product fraction there are a number of different possible operating policies. One for instance, is to operate on a fixed reflux for the whole fraction, producing better than specification material at the beginning and below its specification at the end of the fraction. The fraction change is made when the cumulative receiver contents are at specification.

A second policy would be to choose the reflux ratio that always produces on-specification material, stopping the fraction when the reflux has climbed to some value considered to be 'uneconomic'.

A third policy would be to determine the mathematically optimal profile and take this as the guiding operating principle. The batch distillation is a mathematically fascinating system, and this has lead to a number of studies where the optimum reflux/time or optimum reflux/concentration profile have been located using the Pontryagin Maximum principle.

The conclusions of these studies would appear to be that the constant reflux policy is far from optimal. The policy of changing the reflux ratio to maintain specification would seem to be within a few percent of the theoretical optimum as determined by Pontryagin, as long as an intelligent estimate is made of the maximum reflux ratio at which to change fraction.

Pontryagin studies are not proposed as techniques for determining the optimum policy for each column. They require considerable calculation and are difficult to implement, because often the optimum profile is linked to a difficult-to-measure variable in the system, and they need a microcomputer to carry out the optimal policy. The studies, however, have been useful in quantifying the optimal policies and the magnitude of the losses involved by operating a simpler but more easily implemented policy.

The constant reflux ratio policy is the simplest and easiest to implement thus, for very small systems it may be overall cheaper than installing reflux control equipment to maintain constant product quality to reduce operating costs.

Further reflux ratio operating policies have to be made for the intermediate fractions. The general rule of thumb is to take off the intermediate fraction quickly at a comparatively low and constant reflux ratio.

Related to the intermediate reflux policy is the intermediate fraction recycle policy. Should comparatively large intermediate fractions be taken at low reflux ratios, or would the quantity recycled be best minimized by having high reflux ratios for the product cut-off, and intermediate fractions.

To determine the true optimum for such a problem is extremely difficult – and for a small batch system entirely unjustified. However some attempt must be made to find suitable operating points involving low cost and stable operation. The design method discussed later will indicate how this can be tackled.

5.5c COLUMN FEED POLICY

A batch column will have one or more feed mixtures plus various intermediate fractions to be recycled. The various feeds should not be simply mixed together to provide the column with a single feed, but the 'most dilute' mixture should be charged to the boiler, and the others added when the boiler composition equals the composition of the charges to be fed; this is only good thermodynamic sense!

5.5d RESULTING PROFILES

Reflux ratios change with time and fraction, boiler temperatures rise, corresponding boiler steam temperatures rise – if necessary to the maximum – to maintain column loadings, and then if necessary the column total pressure is reduced to maintain maximum column loading.

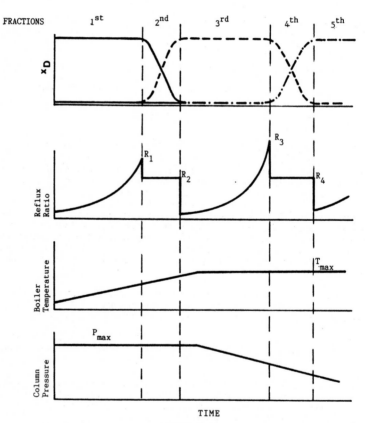

TIME

FIGURE 5.8
TIME PROFILES OVER A BATCH DISTILLATION

The total picture is shown on Figure 5.8. It is not simple: there are many decisions that the operator, or the designer before him, must make before the column is designed or operated.

5.6 BATCH PLANT DESIGN

5.6a THE OBJECTIVE

We have a quantity of mixture to be separated into its components to a specified degree of purity within a given cycle time. We must specify equipment that will achieve this cost-effectively.

The equipment must be specified:

- the number of trays or height of packing
- the column or tray details
- the column diameter
- the condenser area
- the boiler dimensions, its internal and external heat transfer area
- the receiver dimensions

and the appropriate operating policy defined:

- product fractions reflux operating policy - end reflux ratio for
  each product fraction
- intermediate fraction reflux operating policy - reflux ratio for
  each fraction
- systems pressure and pressure profile policy

so that the batch can be distilled within the batch time available.

5.6b THE DESIGN

Preliminary Order of Magnitude Calculations

As with all designs, the first stage is to carry out a few short-cut calculations to appreciate the magnitude of the problem before getting involved in the detail of the design calculation methods or computer models.

What are the boiling points for the components, and the boiling point of the initial and final boiler composition? What are maximum allowable temperatures to ensure there is no thermal decomposition? Answer to these questions allow one to decide what the start and end pressures of the distillation have to be, what the condenser coolant temperature levels required are, and what steam levels should be used for the boiler.

Looking at the charge volume and composition will indicate the size of the boiler and the recovery system, and the boiling point differences between the various components will indicate the magnitude of difficulty of separation.

From this, a very rough estimate of the number of plates and possible reflux ratios can be made, and thus vapour velocities, column diameters, heat fluxes and required areas can be approximately estimated.

At this point we are in a position to suggest a plant configuration that could be used for the first trial simulation. Hopefully, this study has been eased by a data bank that has supplied us with boiling point, densities, latent heat and possibly a few flash calculations.

5.6c DETAILED DESIGN BY HAND

Of course, our major 'guess' has been the reflux ratio required. We can make our estimate more precisely by taking a number of composition situations that we expect during the distillation, with particular emphasis on the end point of each product fraction, and get a series of reflux ratios vs number of plates for each critical situation. From this set a number of plates in the column can be decided upon: then, by repeating a series of snap-shot situations during the distillation, the various reflux ratios required can be determined, and from this a reflux ratio profile built up. This profile can then be used in a stepwise fashion to determine the vapour rate required to achieve the required separation within the allowed batch time, using the Smoker method to perform the integration.

The estimate of number of plates and reflux ratio can be made using the McCabe-Thiele diagram, or the Fenske short-cut equations.

The whole procedure is very tedious and inaccurate, but it is often adequate when very small equipment is involved.

5.6d DETAILED DESIGN BY COMPUTER

To make any reasonably professional attempt at a batch distillation design it is necessary to proceed after the preliminary order-or-magnitude calculation with a computer simulation. Since most batch distillation design packages are based upon simulation of a proposed design, it is necessary to carry out a preliminary order of magnitude calculation to arrive at a rough design to start the simulation study. Although it is conceivable that this be done by repeated trials with the design package, it is advisable to use the initial rough manual estimates, which should always be done before embarking on any CAD study.

A number of batch distillation design programs exist, some company-internal, some belonging to universities and others to software houses. Most packages are based upon the regular hold-up model described in Section 5.4c. The simplest programs consider ideal binary systems; the less simple consider non-ideal binaries; and those of real use and consequence are multicomponent, non-ideal analyses. Such programs are very expensive in terms of computer time, and generally restricted to a simulation of one fraction at a constant reflux ratio.

One program notably different from this treatment is the EURECHA program BATCH which uses the pseudo-steady-state assumption concerning the hold-up to reduce computer time and concentrates more on the behaviour of the total system over the total cycle rather than on a distillation calculation of a single fraction. This BATCH program is available for student use, and the example given at the end of this chapter is an example of use of this program. The program manual is given as Appendix 4.

5.7 THE PROGRAM BATCH

The development of the batch program is an interesting example of the development of software to solve a specific problem, and thus a short description in included here. The problem of batch distillation design is that hand methods are inadequate, yet the classical computer methods are too expensive to be used to design small batch equipment. A design tool was needed that was considerably better then the hand methods but not as thorough as the classical methods, and yet still able to handle multicomponent non-ideal systems.

Naturally the first step is the pseudo steady state assumption for the hold-up. Then some very robust form of convergence is necessary. As previous chapters have pointed out, multicomponent distillation convergence techniques are far from robust and, with them all, occasional 'non-convergence' situations exist. Bisection is the most robust form of convergence method, but it can only be applied to one dimensional systems. However, it is generally possible to convert a multicomponent batch distillation convergence into a one-dimensional problem by the following steps:

(1) Reduce the problem to a ternary distillation;

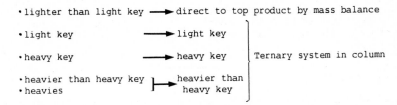

(2) Now the 'heavier than heavy key' (the 3rd component) will not reach the top of the column in significant quantities because the distillation is aimed at keeping the heavy key out of the tops. Hence, effectively all the heavier than heavy key leaving the boiler will be returned to it in the returned liquid reflux.

(3) Hence, a plate-to-plate calculation can be made with only one unknown ($x_{D,1}$) which is necessary to calculate the liquid returning to the boiler (see Figure 5.9). $x_{D,1}$ must lie between 0 and 1.0 and so it can reliably be located by bisection.

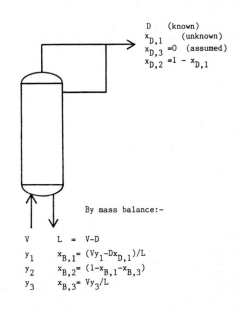

D   (known)
$x_{D,1}$ (unknown)
$x_{D,3} = 0$ (assumed)
$x_{D,2} = 1 - x_{D,1}$

By mass balance:-

V	L = V−D
y_1	$x_{B,1} = (Vy_1 - Dx_{D,1})/L$
y_2	$x_{B,2} = (1 - x_{B,1} - x_{B,3})$
y_3	$x_{B,3} = Vy_3/L$

Hence there is only one unknown:- $x_{D,1}$

FIGURE 5.9

CONVERGENCE OF THE TERNARY COLUMN CALCULATION INVOLVES ONLY ONE UNKNOWN

Hence by reducing the system to a ternary, with a single unknown, further computer time is saved and robustness is improved.

The program was extended by adding heat transfer in the boiler and condenser,
and column pressure drop calculated. This enables control loops to be introduced that act simultaneously with the integration to maintain column loadings
and product specifications by controlling boiler heating medium temperature and
reflux ratio.

The program also considers the different fractions making up the whole cycle,
which enables it to make final a report that shows the cumulative performance of the whole cycle, including total energy use.

The program now fulfils the design requirement in that it is a program   that
considers the whole system, but at a level commensurate with the size of batch
distillation equipment.

The resulting program took much development before it would run reliably. Almost by definition, the introduction of any short-cut method brings problems
with it, because by its very nature it is an abbreviation of the complete picture and is therefore valid over a much more restricted range. Steady development has removed all observable difficulties (some by allowing the engineer to
intervene during the calculation), and the resulting package is used relatively
easily and economically by students and companies.

The results from the program have been compared with experimental results in a
special series of tests carried out in a 500mm diameter industrial column, and
the predicted and experimental composition time profiles compared. The two
profiles were in good agreement and discepancies were within the accuracy with
which the packing HETP was known. The major difference being that the
simulated results gave instantaneous product composition shifts at each change
of reflux ratio, whereas the product composition changed more slowly, over a 15
minute, period with the experimental results. This difference is because of the
pseudo-steady state assumption which does not represent the dynamic behaviour
of the column. The effect of these dynamic differences on the overall profiles
and end compositions was negligable.

5.8 AN EXAMPLE

Design a plant to handle 5 m^3 of benzene/hexane/oil mixture (mf 0.5, 0.3, 0.2)
with a batch cycle time of 11 hours. Product specifications are:

        Benzene   95%
        Hexane   95%
        Oil   90%

To design this plant, we must specify:

- boiler size and heat transfer area
- condenser size
- column diameter, height and internals
- terminal reflux ratio for each product fraction
- constant reflux ratio for each intermediate fraction
- bottoms composition criteria for stripping the intermediate
  fraction.

Many of these variables are highly interactive: reflux ratio/number of
plates/heat transfer areas are related, thus changes in one affects the others.

The basic design procedure is:

(1) Get rough sizes by hand calculation

(2) Use the program to evolve a design that will achieve a distillation with a
    a required cycle time
(3) Investigate modifications to the system defined under (2) to reduce costs.

This last stage can be done by some evolutionary stepwise design, or by a sensitivity study. To attempt to improve conditions by employing some form of
optimization procedure is unlikely to succeed.

STAGE 1

A $5m^3$ charge would require a boiler volume of approximately $6m^3$, with an L/D of
1.5:1.

$$\frac{\pi}{4} d^2 \ 1.5 \ d = 6$$

$$\therefore \ d \simeq 1.7 \ m \quad \text{and} \quad L \simeq 2.5 \ m$$

with a density of 800 kg/m , and molecular weight of approximately 80, there
are about

$$0.8 \ \frac{5 \times 800}{80} \simeq 40 \ \text{kmol/batch overhead product}$$

of product to distil over. The boiling point differences between the
benzene and n-hexane is about 11 K. This indicates a medium difficulty separation, which might require 8 plates and a reflux ratio of 2 to 1.

With this reflux ratio, the hourly loads will be about

$$\frac{3 \times 40}{11} \simeq 10 \ \text{kmol/h}$$

Assuming an 0.3 m/s loading in the column, we can estimate a column diameter;

$$\text{Vapour flow} = 10 \ \frac{353}{273} \times \frac{22.4}{3600} \simeq 0.07 \ m^3/s$$

Hence the column will have to be;

$$\sqrt{\frac{0.07 \times 4}{0.3 \times \pi}} \simeq 0.5 \ m \ \text{diameter}$$

Boiler:

With a latent heat of about 30000 kJ/kmol, the boiler heat load is

$$\simeq \frac{30\ 000 \times 10}{3600} = 83 \ kW$$

With a boiler overall heat transfer coefficient of about 2 kW/m^2K and a temperature difference of about 50 K, a steam temperature of 130 C would be required and the heat transfer required for our boiler will be about

$$\frac{83}{2 \times 50} \simeq 0.8 \ m^2$$

this can be easily supplied by the boiler jacket.

Thus for our first trial we could try with only the jacket providing the heat
transfer area for the boiler. The condenser may have a similar heat transfer
coefficient and a temperature driving force of again 50 K, so giving a
condenser area of about 6 m^2.

STAGE 2

The first simulation took the equipment dimensions estimated above to obtain a first appreciation of the problem.

By studying the output of the simulation it was evident that the steam temperature was too low, the condenser and boiler were too small, and there were not enough plates (25% of the material had to be recycled as intermediate fractions).

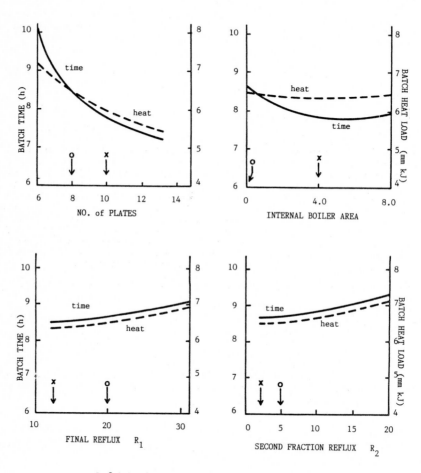

o Origional design before sensitivity study

x Selected design after sensitivity study

FIGURE 5.10
SENSITIVITY STUDY TO IMPROVE THE BATCH DESIGN

A second simulation with equipment defined as:

> Boiler 1.75 m dia x 2.5 m
> Condenser 8.0 m^2
> Column 0.5 m dia x 8 m high   (8 plates)
> Steam temperature 150°C

resulted in an operable design, with a cycle time of 8.6 h. 10% recycled, and a heat consumption per cycle of 6.7 mm kJ. This can therefore be considered to be the end of Stage 2.

STAGE 3

Although the above is an operable design, it is the duty of the designer to propose a system that operates at close to minimum costs. The more significant design variables should therefore be investigated and chosen somewhat more carefully.

From studying the simulation results, the more significant design variables appeared to be:

(1) Number of plates.
(2) Heat transfer area in boiler.
(3) Maximum allowed reflux ratio at the end of the first fraction before switching to the first intermediate fraction.
(4) The reflux ratio of the first intermediate fraction.

A further 4 computer runs were made (a further 12 simulations) in order to determine the sensitivity of the system to the 4 variables. Theoretically the objective function for this sensitivity study should be Discounted Total Costs. However, for batch equipment design this is rather over-sophisticated. A good designer would be content to make a judgement based on a compromise between totalheat/batchandbatch time. The sensitivity of both of these to the four variables mentioned above is shown on Figure 5.10. This figure also shows the base point about which the sensitivity was carried out, (marked with a circle) and a subjectively judged better operating point (marked with an X).

The results of the 6th computer run at this point can be summarised as follows:

> Boiler 1.75 m dia x 2.75 m (4.0 m additional coil area)
> Condenser 8.0 m^2
> Column 0.5 m dia x 10 m high   (10 plates)
> Batch time 6.5 hours
> Heat required/batch 5.6 mm kJ.

Since there are 11 hours available for a batch, the whole equipment can be scaled down linearly:

FINAL DESIGN

> Boiler 1.75 m dia x 2.75 m (2.0 m additional coil area)
> Condenser 5.0 m^2
> Column 0.4 m dia x 10 m high   (10 plates)
> Batch time 10.5 hours
> Heat required/batch 5.6 mm kJ.

The total computer time required for this design was 300 CPU seconds; it required 7 computer runs in which a total of 14 simulations were carried out. This single sensitivity analysis achieved a 15% energy saving and a 23% increase in throughput. It would have been possible to go further – carry out a further sensitivity analysis or look at some of the other design variables, or introduce costs to balance capital and running costs properly - but from the shape of the sensitivity curves there is likely to be less than 5% savings in a further effort. It is likely that for small batch equipment this single sensitivity study is adequate to achieve a design.

5.9 FURTHER READING

Billet, R., 1973. Industrielle Destillation, Verlag Chemie, Weinheim.
                    - contains a useful chapter on batch distillation.

King, C.J., 1980. Separation Processes, McGraw-Hill.

Perry R.H. and Chiltern, C.H., 1973. Chemical Engineers' Handbook, McGraw-Hill.
                    - a rare survey of batch distillation calculation procedures.

5.10 QUESTIONS TO CHAPTER 5

(1) What are the advantages of batch distillation?

(2) Why is batch distillation design a more difficult design problem than continuous distillation?

(3) Why is liquid hold-up important?

(4) What advantages are obtained by using a pseudo-steady state calculation approach?

(5) What factors enter into the selection of reflux ratio policy?

(6) What design methods are available for batch distillation systems?

Chapter 6

DISTILLATION SYSTEMS

Distillation design does not begin and end with a single column. The designer should consider the whole distillation system: the boiler, condenser, the heat recovery and controllability. When a number of separations are required, the total distillation section should be considered together: the separation sequence and integrated operation in terms of heat recovery and controllability.

6.1 ANCILLARIES

Figure 6.1 is a flowsheet of a total distillation system for the separation of a mixture into two products.

Besides the column, which has been the major concern of this book, we have the boiler, condenser, feed preheater, and reflux return system. Control will be discussed separately.

6.1a THE BOILER

The task of the boiler is to vaporize some of the bottom products to provide vapour for the separation. It is heat exchange equipment and comparatively expensive. The designer should have the following objectives in mind:

(a) keep the equipment cost low (maintain good heat transfer coefficients);

(b) keep the heat exchange surface as clean as possible (prevention of deposits of degraded products);

(c) keep residence times and wall temperatures low if product thermal degradation is likely;

(d) contribute to the separation required.

There are basically 3 types of boiler in normal use (as shown in Figure 6.2).

The kettle-type boiler (Type A) consists of a submerged tube bundle in a horizontal shell half filled with liquid. A baffle arrangement at the liquid exit end allows for vapour disengagement and enables a level control sensor to operate reliably. The large capacity of a boiler gives stability, particularly if a number of columns are involved, and large vapour disengagement area prevents entrainment of liquid with the vapour. The low vapour-disengaging velocity may also be advantageous in foaming systems. Since the exit liquid flows after part vaporisation, there is an enrichment equivalent to 1 theoretical plate. The horizontal layout also means that the height is reduced, and so the total column and structure height can be correspondingly reduced.

This boiler type is not suitable when the boiling products are thermally unstable. The long residence time also increases the control problem in the column itself (although it reduces the control problem in a series of columns).

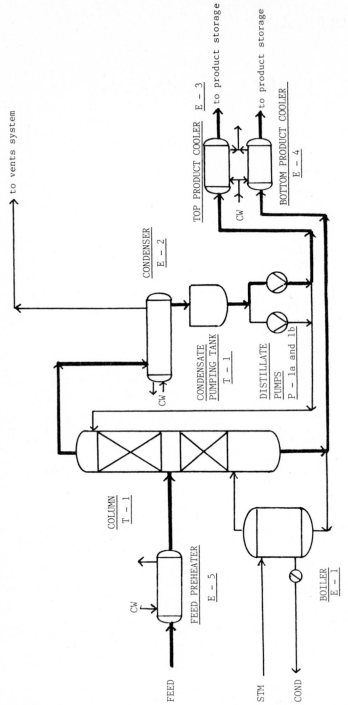

FIGURE 6.1

FLOWSHEET OF A DISTILLATION COLUMN WITH ANCILLARIES

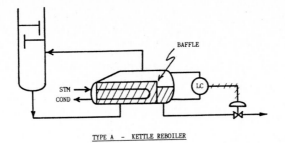

TYPE A - KETTLE REBOILER

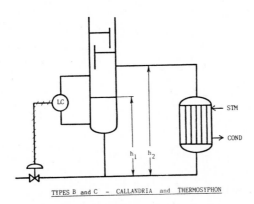

TYPES B and C - CALLANDRIA and THERMOSYPHON

FIGURE 6.2
BOILER TYPES IN NORMAL USE

The callandria once-through type of boiler (Type B) has a lower liquid capacity and so is more suitable for thermally unstable systems. The liquid contents are held up in the column sump. This, together with the vertical installation of the callandria makes it essential that the column be some meters off the ground. The 'once-through' nature of the flow means that the vapour composition is the same as the bottom product composition. There is no enrichment given by the boiler.

Type C is a thermo-syphon callandria boiler in which the liquid levels are so adjusted that the liquid/vapour mixture in the callandria, with its lower density than the quiescent liquid in the column sump, induces a circulation - following the principle of the airlift pump. This circulation increases the fluid velocities in the callandria, which improve the heat transfer coefficient and reduce fouling. The vapour has a different composition to the liquid, and at high circulation rates this gives a benefit approaching 1 theoretical plate.

The thermosyphon has many good properties, and is particularly attractive if there is a chance of thermal instability. In practice there are more difficul- ties in operating at the enhanced heat transfer rates than calculation on paper would suggest. The pressure balances and the liquid levels must be carefully controlled to obtain maximum circulation rate. Unless this is correctly done, the performance does not achieve the design rating.

The correct design requires the hydrostatic pressure of the liquid feed, h_1 on Figure 6.2, to be greater than that of the two-phase mixture in the boiler and vapour take off, h_2 .

This pressure difference induces a flow, whose rate depends on two-phase flow
pressure drop correlations, and this flow determines the heat transfer coefficient, which in turn determines the vapour fraction and hence the two-phase
density. Iterative solution by computer is unavoidable.

This is not the place to discuss the design of heat exchange equipment.
Predictions of the two heat transfer coefficients is done using normal methods.
It should be noted, however, that the boiling film coefficient reaches a maximum heat flux, which is limited by the vapour lining the surface and preventing
further heat transfer even if the heating medium temperature is increased - all
that happens is that wall temperature increase and the effective area
decreases. Boiler areas are therefore designed for this optimum heat flux.
Correlations are available in the literature relating this maximum heat flux to
liquid physical properties (see Perry). A check should be made on the resulting
wall temperature, and if this is considered detrimental to the bottom product,
lower temperatures used and the design made for less than this optimum heat
flux.

When the boiler contents are subject to thermal degradation, it is important to minimise the boiler hold-up and the contact time between the boiler
contents and the heating surfaces. Equipment with suitable characteristics for
this situation is the falling film (or wiped film) evaporator. Contact times
are very low,and the total boiler hold-up can be designed to be very small.
Thermal decomposition is of primary importance in vacuum distillation and so
this equipment is described in more detail whilst discussing vacuum
distillation systems in Section 6.2c.

When the bottom product is waste water, it is quite customary to inject
live steam directly into the column sump. This saves on the capital cost of the
boiler, the only disadvantage being the increase in bottom product (which, if
it requires treatment, involves a great volume to be treated) and lower
concentrations, which increase the separation problem marginally. This effect
can be investigated by considering the steam to be a vapour feed in the column,
and this will effect the mass balance and operating lines, which in turn will
enable the effect on the separation to be calculated.

It is common practice to use this 'live steam'; clearly, these disadvantages
cannot be not very significant.

6.1b THE CONDENSER

The task of the condenser is to condense sufficient vapour leaving the top of
the column to provide sufficient reflux for the separation. If a liquid top
product is required, the condenser condenses the total vapour stream. In situations where the top product is a wide boiling mixture, the condenser condenses
the reflux, plus the heavier components, to form the liquid product, and the
non-condensed fraction forms a second vapour product. Such an arrangement is
common in the oil industry.

Partial condensation has the advantage that the non-condensed product is
enriched in the light component. The condenser is providing a further theoretical plate of separation. When condensation is total, and part is returned as
reflux, this enhanced separation is lost.

The difficulty with partial condensation is that the reflux rate is controlled
by regulating the cooler flow through the condenser. This provides a slow control system because of the thermal capacities involved; it is also very difficult because of the non-linear relationship between coolant flow and performance in an economically designed heat exchanger. This is discussed in more
detail in Chapter 8.

Condensers can be be mounted above the column, or near ground level. For large columns, a water-cooled condenser becomes a very heavy piece of equipment, and it is more economic to pipe the vapours to ground level and mount the condenser at ground level than build a sufficiently strong supporting structure above the column to hold the condenser.

If the condenser can be mounted above the column, it may be possible to lead the reflux back into the column by gravity. A ground level condenser re-quires a pump to return the reflux to the top of the column. The pump requires the throughput to be controlled and this requires a pumping tank with a level controller; the working pump requires a standby spare, and so a ground mounted condenser requires considerably more hardware than one mounted above the col-umn. Figure 6.3 compares the different arrangements.

Ground-mounted condenser is a misnomer: the pump is mounted on the ground. The liquid head above it must correspond to the net positive suction head (NPSH) required for the pump, and this determines the pumping tank height. Above this tank is the condenser, on a structure some 5 meters above the ground.

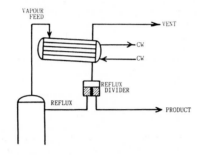

a)   CONDENSER MOUNTED ABOVE COLUMN

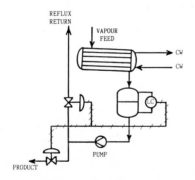

b)   GROUND-MOUNTED CONDENSER

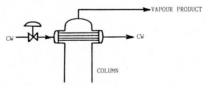

c)   PARTIAL CONDENSER or DEPHLEGMATOR

FIGURE 6.3
DIFFERENT CONDENSER INSTALLATIONS

The partial condensers, mounted above the column in such a way that the condensed vapours rain back into the column is called a dephlegmator. It is the simplest form of installation, as long as the condenser is not too big. Its control is poor and there is no measure of the reflux returned to the column. However if the top vapours are wide boiling, and the condensate needs to be returned and the lights taken as a vapour product, and the cooling water temperature is in the right range, then the dephlegmator is ideal.

Condensers can either be water (or other liquid cooling medium) cooled, or air-cooled. The normal considerations apply for making the choice: air cooling has high capital costs, but low running costs; it becomes attractive over cooling water when the equipment becomes small, i.e. when condensing temperatures are high. Each firm has its own individual studies to determine the 'break point' and this is used as a rule of thumb for all its designs. The 'break point' used to be 100 °C, but it changes steadily as improvements are made in air cooler design. Figures of 60 °C may now sometimes be justifiable.

Air coolers are mounted high, usually above the columns, because they are lighter and require free space to prevent recirculation of used air.

The sizing of condensers is done using correlations for the two films, the condensing film coefficient usually being the better of the two.

Shell and tube condensers are usually mounted horizontally with the condensed vapours in the shell side. This allows the water to be in the tubes, where high velocities can be obtained to improve the film coefficient. The vapours are feed into one end of the shell, and the vent is taken from the other. Inert gases and non-condensables then leave the condenser with minimum content of condensables.

Since condensers are usually run with fixed cooling water flow, the variable load results in a variable area being used for the condensation. This can be easily observed by the temperature of the shell of the condenser. There is an abrupt temperature change along the condenser denoting where all the vapour has condensed. When the limit reaches the vent pipe, the condenser is overloaded.

6.1c FEED PREHEATERS AND PRODUCT COOLERS

These exchanges serve to adjust the temperatures of the respective streams to their required values.

The feed temperature to the column is chosen to prevent column capacity restrictions within the stripping section of the column, where extra vapour has to be supplied to heat up the feed to its boiling point. Rather than increase the diameter of the whole column, it may be more economic to preheat the feed, thus reducing the load in the stripping section. The feed conditions therefore affect the L and V flows, both above and below the feed, which in turn determine a number of stages and loadings in each column half. The feed temperature is specified to obtain the best total column design.

The specification of the product temperatures is determined by the need to have a specific temperature for the next processing step, or to have a safe storage temperature (considering transportation, vapour pressures, inflammability etc).

These duties are pure heat exchange duties, and normal criteria for choice of exchanger and design apply. The feed preheating and product cooling are obviously possibilities for heat integration. This is discussed later in this chapter.

6.1d REFLUX SYSTEMS AND PRODUCT REMOVAL

A fraction of the condensed liquid must be returned as reflux. This quantity
has to be known, measured and controlled. The most positive way of achieving
this is to collect all the condensate, and split it into two flows, which can
be measured and controlled with orifice meters (for example) and control
valves.

Such a system requires an appreciable pressure drop, and a permanently full
pipework system. To achieve this, it is necessary to have a condensate tank,
pump (and install spare) and a level controller on the tank (see Figure 6.4
for a possible arrangement for a reflux system).

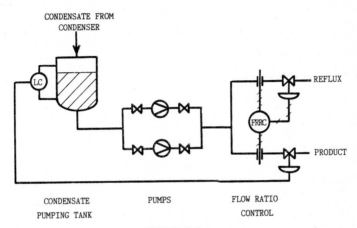

FIGURE 6.4
REFLUX CONTROL SYSTEM

There are of course simpler methods of achieving a known adjustable split of
liquid from a condenser. A very simple method is the 'reflux divider', which is
a rotatable slotted bucket in a small divided tank (see Figure 6.5).

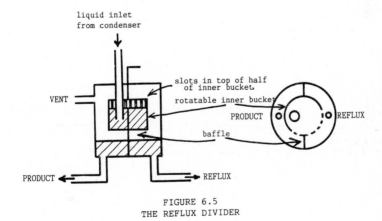

FIGURE 6.5
THE REFLUX DIVIDER

The position of the inner slotted bucket determines how many slots feed each half of the outer container, and so control the split of the condensate flow between reflux and product.

This equipment is installed and functions on old plants, but it is a frequent source of trouble. Often the correct flow rates are not maintained. The problem is that there is very little head to work with, since it is gravity fed. The outer tank may fill completely if there is a back up in the system or a problem with the vent balancing.

However well thought out a design, experience shows that reflux dividers are unreliable. Hence, they cannot be recommended for installation in a modern plant. Nevertheless they are worth mentioning from the design point of view; they offer a low cost clever solution to replace a complex system. However, the design is 'too clever', and deviations from the ideal design conditions cause malfunction and unreliability - two important design lessons in one. It is, however, worth considering by the entrepreneur who wishes to bring his ideas to fruition with a low cost plant in his back-garden.

Generally, the bottom product is a boiling liquid that has to be removed from the boiler or column sump to maintain a fixed bottoms level. This is generally achieved by level control in the sump, or behind the baffle in a kettle reboiler. The product is then cooled and pumped to product storage (see Figure 6.6).

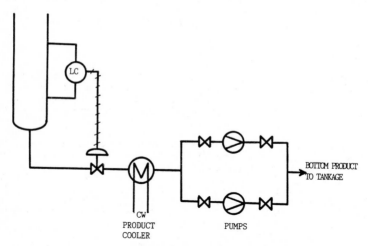

FIGURE 6.6
BOTTOM REMOVAL SYSTEM

Whether cooling is done before or after the pump dependents on the NPSH that the pump requires, and what is available from the 'boiling' liquid. The decision depends upon the height difference between the bottoms level and the pump.

Again, cheaper designs are possible. A minimum cost installation consists of a steam trap installed at the level to be controlled. This opens with high density (liquid) and closes in a low density environment. Once open, the liquid is free to flow to the product tanks under its own pressure arising from the pressure drop in the column (see Figure 6.7).

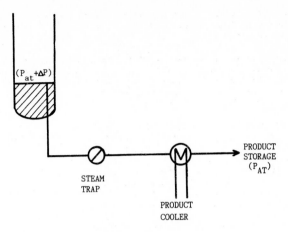

FIGURE 6.7
BOTTOM DISCHARGE BY MEANS OF A STEAM TRAP

Again, these systems have been installed, but they are unreliable: a novel design but again 'too clever'. The steam trap does not have a positive shut-off,
the column pressure drop is sometimes not enough to discharge the liquid. This
results in a trap opening but the level still continuing to increase. Once
more, useful to the entrepreneur, but not to be installed in a complex plant.

6.1e TANKAGE

The feed tank is there to enable a constant feed to the column to be maintained
despite changes in upstream flow rate. This may come from an upstream plant
that does not run at a constant rate identical to the downstream plant rate,
or it may be that the feed is batchwise produced or delivered, and so a tank is
essential.

The tank size is determined by considering the possible fluctuations in the de-
livery of the feed and the required degree of stability required by the feed.
In a chain of distillation columns, a rule of thumb is that there should be a
20 minute hold-up between columns. For delivered feed, the tank volume may be
that required to maintain operation during the longest national holiday (if
this means that deliveries are stopped but the distillation plant is expected
to keep running). Methods of optimal sizing these tanks involve economic order
quantity calculations, reliability and control studies and stochastic
simulation – should the size be judged important enough to merit a deeper
study. These techniques are, however, outside the scope of this text.

The products can either be run directly into product tankage, or into smaller
receivers which are quality controlled before being discharged to tankage.

The more modern approach is to run directly to product tankage with a continuous analysis of the product run-off, and a warning system for when specification is not achieved. The danger with this system is that off-spec product does
reach the tankage (e.g. when the analysis instrument fails) and then the whole
stored product is off specification, and must be re-worked. This situation requires two product tanks, in order to prevent a total product stoppage and enable the off-specification tank to be reprocessed.

The alternative system is to collect smaller receivers (e.g. shift-receivers)
of a product, have the quality controlled, and discharge this to a single

FIGURE 6.8
PETROCHEMICAL DISTILLATION PLANT: Free-Standing Column, with Boiler, Condenser
and Receivers in the Structure behind the Column

FIGURE 6.9
VACUUM DISTILLATION PLANT: Supported Column with Overhead Condenser and
Falling-film Boiler

116   Distillation Design in Practice

product tank as long as it is on specification. This system requires multiple
small receivers, but just one product tank. It requires more labour, and is
less convenient when the plant is in steady operation. The final choice of system depends upon production rates, and the relative seriousness of the consequences of out-of-specification production.

Product tanks have a further function in smoothing out product quality
fluctuations. To fill drums directly from the condenser would produce a product
quality variation per drum that would probably be unacceptable and difficult to
check. A mixed product tank reduces quality variations and reduces analytical
work. It can be shown for a batch system that the reduction in standard deviation from small batches volume v, by putting them through a mixed tank volume V
is given by:

$$\sigma = \frac{\sigma_o^2}{1 + 2\frac{V}{v}} \tag{6.1}$$

where σ is the standard deviation of the product from the tank
and σ_o is the standard deviation of the product to the tank.

6.1f PLANT LAYOUT

The two photographs of distillation units (Figures 6.8 and 6.9) give some idea
of how the column and its ancillaries can be laid out in practice. Figure 6.8
shows free-standing columns, typical of tall, large diameter columns, with the
condenser mounted at ground level. Figure 6.9 shows a layout more suitable for
smaller columns, with the column supported in a structure and the condenser
mounted overhead.

In laying out the distillation section of a plant one attempts to collect all
the columns together, since this produces a more economic supporting structure.
Likewise, it is good to collect heat exchanges together, since a common structure can then be designed. Furthermore, heat exchanges require a free distance
equal to the length of the heat exchanger, in front of the heat exchanger, to
enable the tubes to be withdrawn for cleaning and replacement. This is often
difficult to arrange, but by collecting exchangers together, there are fewer
tube-withdrawal free areas to place on the site and this facilitates layout
design.

Against this principle of collecting similar items together is, of course, the
fact that they are rarely adjacent in the process. There is a limit to which
one is willing to pump process streams long distances, simply to arrive at a
convenient layout. Pressure drop, dead-times, and piping costs dictate that
compromises are sought, where the best solution may be to have more than one
site for each equipment type.

In terms of a design exercise, layout is fascinating. The possibilities are
endless; the criteria a mixture of subjective judgement, calculation and
economics; and constraints in terms of maintenance requirements, safety rules
and regulations abound.

6.2 THE SINGLE COLUMN

A simple distillation column has a single feed which is split into a top and a
bottom product. Depending on the column operation and the column size, these
two products can be, to a greater or lesser extent, a clean separation between
the components in the mixture. Hence for a binary separation we can obtain pure
components (assuming the VLE allows this to be done).

$$A + B \rightarrow A/B$$

For a multicomponent mixture, the single column can still produce only one
clean separation, so one can either produce one component or a clean separation
between two mixtures.

$$A + B + C + D \rightarrow A/B + C + D$$

$$\text{or} \quad A + B + C + D \rightarrow A + B/C + D$$

Hence the rule that, in order to separate an n component mixture into n pure
components, n - 1 columns are necessary.

6.2.a SIDESTREAMS

This rule does not apply if <u>pure</u> products are not required. For a
multicomponent mixture, the column profile shows that different compositions
are most concentrated in the different part of the column. If products are removed at these points (<u>i.e.</u> <u>sidestreams</u> <u>taken</u>), then a number of products may
be obtained which are enriched in different components.

The sidestreams will not be pure components: above the feed, all sidestreams
must contain the lightest products and below the feed all sidestreams must contain the heaviest products, because these components are passing the sidestream
point, and therefore must be present (to fulfill at least the mass balance
requirement). Furthermore, the very removal of the sidestream alters the L/V
composition in the column, thus the enrichment obtained when a non-negligible
quantity of sidestream is removed is less than a profile without sidestreams
would predict. However, depending on relative concentrations and relative product specifications, sidestreams can be useful since each sidestream replaces a
total column.

In the chemical industry, sidestreams are used to improve product qualities by
removing small quantities of light or heavy components that are present. If a
column feed has two major components, and a small quantity of a lighter
impurity, and a small quantity of heavier impurity, then, if the products are
taken from the first and the last plate of the column, rather than the boiler
and condenser, and only a small purge is taken from the condenser and boiler,
there will be a considerable reduction of these extreme impurities. The system is described in Figure 6.10.

Although the product sidestreams cannot be 'pure' components, depending on relative quantities and relative volatilities, the sidestreams can be within specification. Depending on the number of plates above and below the sidestream,
it is possible to adjust the light and heavy composition to minimize loss of
product.

The quantity of the two purges, reflux and vapour rate, removal of sidestream
as liquid or vapour and the number of plates in the 4 column sections are all
degrees of freedom that the designer has to define. The total system is
straightforward to calculate. The basic equations developed in Chapter 4 still
apply. The VLE equation to determine the vapour compositions over the liquid
remain identical, the mass balance equations alter at each sidestream point,
exactly as they do over the feed point in the simple columns. Figure 6.11 shows
the total flow mass balance equations over the feed, vapour and liquid
sidestream points. Component mass balances can be written down in an equivalent
manner.

This represents such a minor modification to the standard distillation
simulation computer program that most programs are written in a perfectly general way, enabling vapour or liquid sidestreams to be removed from any (or
every) plate.

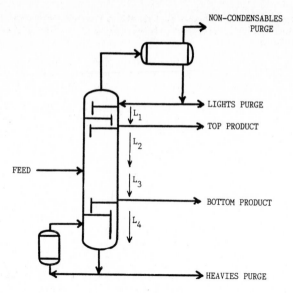

FIGURE 6.10
HEAVY AND LIGHT TRACE IMPURITY REMOVED BY THE USE OF SIDESTREAMS

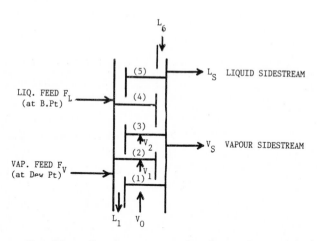

Mass Balance Equations	Heat Balance Equations (CMO)
Plate 1: $V_0 + L_2 + F_V = V_1 + L_1$	$V_1 = V_0 + F_V$
Plate 2: $V_1 + L_3 = V_2 + V_S + L_2$	$V_2 = V_1 - V_S$
Plate 3: $V_2 + L_4 = V_3 + L_3$	$V_3 = V_2$
Plate 4: $V_3 + L_5 + F_L = V_4 + L_4$	$V_4 = V_3$
Plate 5: $V_4 + L_6 = V_5 + L_5 + L_S$	$V_5 = V_4$

FIGURE 6.11
MASS BALANCE EQUATIONS OVER THE FEED AND VAPOUR AND
LIQUID SIDESTREAM POINTS

The design procedure for a column with sidestreams is therefore exactly the same as the design of a single multicomponent column, except there are more degrees of freedom.

A useful start is the mass balance analysis as shown on Figure 6.11, and to calculate by hand the appropriate flows V_O - V_N and L_1 to L_N, which depend on the quantities of components present in the sidestreams and feed.

Having determined values for D, B, L_S and V_S from feed compositions and specifications and suitable L/V ratios in the various sections, a reasonable starting estimate for the 'reflux ratio' can be made and the investigations started.

In the oil refining industry, sidestreams are more frequently used than in the chemical industry, because crude oil consists of many hundreds of components, and the problem is not that of isolating pure components but of getting a product with particular physical properties.

The profile of a crude oil distillation unit (pipestill) contains points where all the product ranges can be tapped off: gas, oil, kerosine, naphtha, LPG. Hence, in the single distillation column, the different products can all be obtained by taking off different sidestreams. Of course, for the light components to reach the top of the column, these must be present in all the sidestreams above the feed, and so the sidestream products, even for these petroleum fractions, must have the light components removed. Hence each sidestream has to be stripped of lights before it yields on-specification product.

The design of such systems remains, in principle, the same as the normal multicomponent system. Simulations must be done with a feed containing many components, many of which are 'pseudo components', but the same distillation program can be used. The removal of sidestreams is done exactly as shown in Figure 6.11.

6.2b MULTIPLE FEEDS

Since the objective of a distillation column is the separation of a mixture into two components - to decrease the total entropy of the two components - it is a thermodynamic nonsense to mix together streams of different concentrations before separating. In the limit, if two streams are available with the correct top and bottom specifications, they should not be mixed together and then separated, but fed to the top and bottom of the distillation column (or not fed at all!). If the two streams do not match the product specification, they should be fed nearer to the top and bottom of the column, respectively, but not mixed together and fed in the middle.

Additional feeds should not be mixed but fed to the column at the appropriate point: where feed compositions match the column profile. Hence, it is quite usual to have more then one feed to a column. Multiple-feed columns are treated in exactly the same way as the single feed column. Equations remain the same, except that there are more feed plates, more changes of mass balance equations and more sets of L/V. The best location of each feed is found in the same way as the location of the single feed.

6.2c VACUUM DISTILLATION

Vacuum distillation is necessary when the boiling point of the material to be vaporised is too high at normal pressures for the material to remain stable. By working at lower pressures the boiling point is reduced, and thermal decomposition can then be avoided. The limit of distillation as a separation process is when vaporisation cannot occur before thermal decomposition occurs, and so by going to lower and lower vacuum, distillation has greater and greater application. Pressures as low as 10 mbar are used in special instances for high boiling unstable substances. At these pressures the distillation process is call 'short path' or 'molecular distillation' because the distance between the boiler and the condenser is of the same order as the mean free path of the molecules.

For anyone designing a vacuum distillation column it very quickly becomes evident that pressure drop is of primary importance. The pressure drop of the commonly used internals is of the same order as the total pressure required in the boiler of a conventional vacuum distillation (e.g. 0.05 bar), meaning that the condenser pressure would have to be zero! Internals have to be chosen on the criterion of pressure drop per theoretical tray. The greater the vacuum, the more sophisticated the internals. As higher vacua are required, falling film empty tubes are necessary to keep the pressure drop low. At even higher vacuum, the 'wiped wall' column –a wetted wall with a rotor – is necessary. For the highest vacuum, molecular distillation equipment, which has the condensing surface only a few centimeters from the heating surface is necessary (see Figure 6.12).

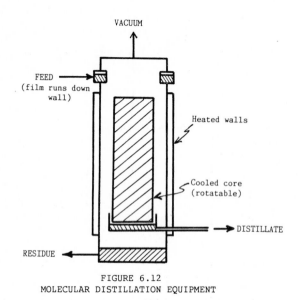

FIGURE 6.12
MOLECULAR DISTILLATION EQUIPMENT

Table 6.1 summarises the working range for the various types of internals. Chapter 7 discusses these internals in more detail.

Because the reason for the need for vacuum distillation is usually associated with thermal decomposition in the boiler, the boiler design of a vacuum column is often non-conventional, in order to reduce residence times and reduce contact time with the hot surfaces. Batch distillations are particularly vunerable because the material stays in the boiler for hours. Falling film evaporators, or wiped film evaporators, are ideal (but expensive) methods of providing vapour for the distillation. The boiler contents are pumped through the falling film evaporator, and vaporisation occurs, the contact between the material and the hot surface being only of the order of seconds. Figure 6.13 shows a typical arrangement. A high circulation rate ensures that the boiler achieves 1 theoretical stage of separation.

TABLE 6.1
OPERATING PRESSURE RANGES FOR DIFFERENT INTERNALS

PRESSURE RANGE (millibar)	SUITABLE INTERNALS
down to 200	plates
100	conventional packing (e.g. Raschig rings)
50	improved packing (e.g. pall rings, saddles)
10	Sulzer-type packings
2	wetted wall column
1	wiped wall column
0.001 to 0.0001	molecular distillation equipment

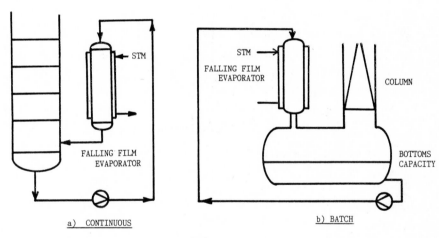

FIGURE 6.13
FALLING FILM EVAPORATOR AS BOILER TO A VACUUM BATCH DISTILLATION

6.3 MULTIPLE COLUMNS

As previously stated, if n pure fractions from a mixture of n components are
required, n - 1 columns must be provided. The use of sidestreams is possible
only when 'impure' fractions are acceptable, or when some of the components are
in such low concentrations that they do not make the sidestream 'impure'.

A distillation train or sequence must be used, with each column making a clean
separation.   Figure 6.14 shows the possibilities for a 3 component mixture A,
B, C (in the sequence of relative volatilities).

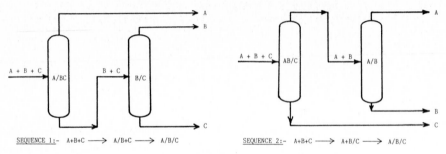

SEQUENCE 1:- A+B+C ⟶ A/B+C ⟶ A/B/C   SEQUENCE 2:- A+B+C ⟶ A+B/C ⟶ A/B/C

FIGURE 6.14
SEPARATION OF A TERNARY TO PURE COMPONENTS

The design problem associated with distillation trains is to determine the best 'order' in which to make the separation. For the ternary example there are basically two ways: for a quaternary there are 6 configurations with 3 columns, and for 5 components there are 24 configurations with 4 columns.

It is not unusual for a mixture to have 5 components, thus the design problem can be appreciable. To choose the optimum from 24 possibilities involving design and costing of 96 columns is no easy matter.

6.3a RIGOROUS METHODS OF COLUMN SEQUENCING

A full enumeration of the possibilities produces the combinatorial requirement of these 96 column designs. However, many of these designs are identical, and this leads to the application of integer optimization methods, which reduce the number of designs necessary.

Figure 6.15 shows the total set of possibilities with the 24 schemes and 96 separations in 4 stages. Each stage represents a further separation. What is clear from this figure is that, although there are 96 separations (4 x 24), these are not all different, since many of the routes are common along much of their paths. The compilation problem can therefore be reduced by employing either branch and bound or dynamic programming, which are integer optimization methods designed to profit from the tree structure of Figure 6.15.

To employ branch and bound, the costs for each full route are calculated by working from left to right through the tree.

Costs at each branch point (node) are stored, in addition to the final cost of the first route. Further routes are calculated, returning to the latest node, so that the second routes require only 2 distillations to be calculated, and the third only 3 and the fourth 2 and so on. The total costs are obtained with only a fraction of the total calculation effort. This is the 'branching' part of the algorithm. The computation can sometimes be further reduced by the bounding part, whereby any calculation is broken off as soon as the total cost-so-far exceeds the lowest yet total cost route. This is originally route 1, but it is updated as better routes are found.

The second possibility is to employ dynamic programming. For this method Figure 6.15 has to be redrawn as Figure 6.16, where for the 4 stages of the separation are shown all the various possible 'states' i.e. A/B + C + D + E is a state, as is A + B/C + D + E. This diagram shows a much simpler situation , with only 15 different states being possible.

The idea of dynamic programming is to reduce the magnitude of the problem by building up the optimum route from optimum sub-routes. Once the optimal route to a state is known, if the overall optimal route goes through that state, it will use the optimal route to that stage. This is a common sense statement which serves to cut out a lot of the analysis of complete roots.

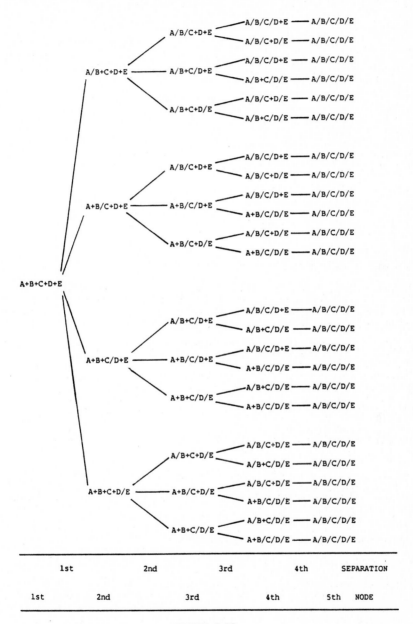

FIGURE 6.15
TOTAL ENUMERATION OF THE 24 ROOTS SEPARATING A 5-COMPONENT MIXTURE

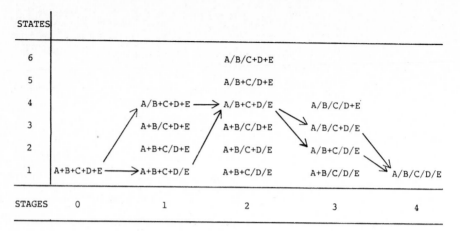

STATES

6	A/B/C+D+E
5	A/B+C/D+E
4	A/B+C+D+E → A/B+C+D/E A/B/C/D+E
3	A+B/C+D+E A+B/C/D+E A/B/C+D/E
2	A+B+C/D+E A+B/C+D/E A/B+C/D/E
1	A+B+C+D+E → A+B+C+D/E A+B+C/D/E A+B/C/D/E A/B/C/D/E

STAGES   0   1   2   3   4

(The arrows correspond to the last example in the text)

FIGURE 6.16
TOTAL ENUMERATION OF THE DIFFERENT STATES WITH 4 STAGES OF THE SEPARATION
OF A 5-COMPONENT MIXTURE

For example, from Figure 6.16 state A/B + C + D/E has 2 routes to it, and 2 routes from it, making a total of 4 that pass through this state. Now by costing the 2 routes to it, the best can be chosen and this single route used in combination with the two routes from it. There is no point in evaluating the other two route combinations because they must be worse. Hence, this calculation is reduced by 50%.

Dynamic programming calculations progress forwards (or backwards) and store the best root to each state, and its cost. The calculation moves stage by stage through the tree (not route by route as with branch and bound) and all the routes are built up using the optimal route so far. Only when the whole tree is finished are there any results, but the total number of complete routes evaluated has been drastically reduced, and the optimal one is found.

These two methods have both been proposed for selecting optimal distillation train designs. To complete the picture they can be programmed so that the complete enumeration is done automatically and the computer program simply prints out the best route (or a selection of the best routes).

Whether such fully automatic methods will be acceptable to designers is another matter. The problem does not arise often enough for a good distillation program with accurate, company accepted costing to be included, and there are many other factors that influence column sequences that would have to be programmed into a fully automatic method.

6.3b   HEURISTIC METHODS OF COLUMN SEQUENCING

Rather than fully evaluate all the possible routes, a further possibility is to use a set of heuristic rules which generally produce a good sequence. A number of studies have been undertaken, and sets of rules have been devised as a result of carrying out many case-studies. For instance, those of Nishimura are:

(1) separations where the relative volatility of the key component is close to unity should be performed in the absence of non-key components;

(2) sequences that remove the components one by one in the column overheads should be preferred;

(3) sequences that give a more equimolar division between the distillate and the bottom products should be favoured;

(4) separations involving very high specific recoveries should be left until last;

(5) for feed mixtures that contain one component in excess, a sequence that removes this component as early as possible should be favoured.

Clearly, many of these rules are contradictory. The problem is too complex because of the large number of variables (volatilities, specifications, feed concentrations) to produce more satisfactory generalisations. However, these rules are a great help because they enable many alternatives to be discarded and a reduced set to be brought forward for further investigation.

There are considerable differences as regards the costs of different sequences, as shown in a paper by Freshwater and Hendry. Under some circumstances, total cost variation (investment plus operating costs) of 50% between the best and the worst sequence were found possible.

6.3c A DESIGN METHOD

In addition to our rigorous method and our heuristic method, each separation problem has its own peculiarities. The feed may be dirty, or it may have boiler decomposition problems, or product fouling or corrosion problems as a result of thermal degradation.

Hence, in such a case, more rules appear; it is better to remove the heavy component first, taking all the other products as distillate. This means that only one column sees a dirty duty and is subject to problems of thermal degradation, fouling or corrosion.

Since a reliable automatic sequencing in program is most unlikely to be available to the design engineer faced with a sequencing problem, how should he find a solution?

Firstly, there are specific rules for this particular system, and there are usually constraints that must be fulfilled. Then the heuristic rules suggest a number of different possibilities – using these rules rather to exclude unlikely sequences than choose the best. Hence, the design can evolve a number of different possibilities that should be good.

The tree structure according to Figures 6.15 and 6.16 can then be constructed, including only those chosen structures and, using the principles of dynamic programming or branch and bound, the optimum can be found using the minimum number of calculations. There is no need to have these logical methods programmed; they are equally useful to the designer who is considering his alternatives off-line.

Hence, the practical approach for the designer in determining his distillation sequence is to make use of all the rules and restrictions available to produce a sub-set of all the possible sequences, use a normal distillation program to evaluate these specific separations involved, and construct these into total routes, making use of the tree structures involved.

6.4 HEAT INTEGRATION

The distillation column is a heavy energy consumer. Feeds must be heated up to
distillation temperature, and vapour produced by boiling bottom products to
produce the counter current flow for separation.

Likewise, it has a high demand for cooling. The vapours must be condensed and
the products cooled. There is great scope for heat integration to achieve the
required separations with reduced energy demands.

6.4a THE SINGLE COLUMN

The most obvious source of heat recovery is the heat from the bottom product,
which can be used to preheat the feed to the column. This is a perfectly
straightforward heat exchange duty. There is often insufficient heat available
in the smaller bottom product stream, and this requires a further preheater to
be installed after the heat interchanger (see Figure 6.17). This is not a bad
thing from the control consideration, because it enables the external heat
source to compensate for boiler flow changes that otherwise could induce
instabilities.

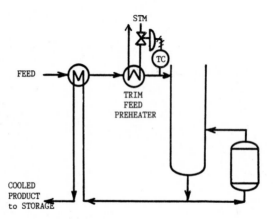

FIGURE 6.17
HEAT RECOVERY FROM BOTTOMS PRODUCT BY HEAT EXCHANGE WITH FEED

A   common   use   of   distillation   in the chemical industry is in the
absorption/distillation sequence in which a product is recovered from a reactor
exit stream. Here, large quantities of solvent can be circulating, and this
produces a sensible heat load considerably greater than one would expect from
looking at the sensible heats associated with the product itself, since these
solvent recycle rates can be 5 or more times the product flow rate. A typical
recovery scheme is that given in Figure 6.18.

The most renowned feed pre heat/heat recovery system is, of course, the pipe-
still feed which must be heated from 10°C to 300°C before being fed to the column. The sensible heat required is enormous. A 100,000 barrel/day refinery
would have a pipe-still feed heat requirement of 50,000 kW.

The refinery has this one large flow requiring heat, and many smaller flows
giving heat (product coolers, condensers, etc). There is enormous incentive to
re-use all possible heat in pre-heating the pipe-still feed.

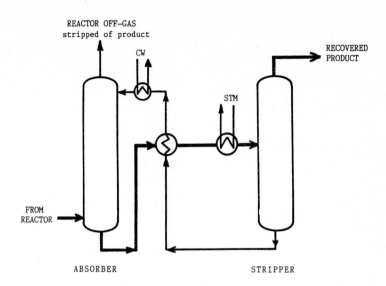

FIGURE 6.18
HEAT RECOVERY IN AN ABSORPTION/DISTILLATION PRODUCT RECOVERY SYSTEM

The traditional way in which this design is achieved is to present the pipe-still feed at an enthalpy/temperature plot, and provide each hot stream again as enthalpy/temperature plot. The configuration of the heat exchanges to achieve the heat exchange is then determined by manipulating the hot stream positions on the pipe-still feed plot until a satisfactory arrangement has been found. See Figure 6.19, which shows the single cold stream A being heated by hot streams B, C, D, E and F.

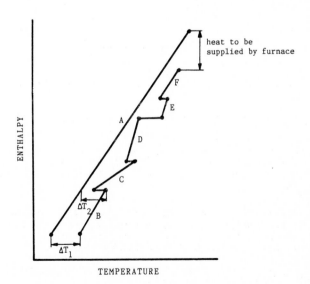

FIGURE 6.19
ENTHALPY TEMPERATURE CHART FOR A SYSTEM OF HEAT EXCHANGES

This chart shows the position of the individual heat exchangers in the sequence; it shows the temperature driving forces, and it shows the quantity of heat that has to be supplied by the furnace to achieve the required feed temperature. The incentive for heat recovery now requires that more complicated systems are investigated, whereby stream splitting is used to improve recovery, and more than one single hot stream is brought into the study.

This becomes a vast combinatoric study which is best solved by a mixture of searching and heuristic rules, A detailed discussion is outside the scope of this text.

A method that has been quite successful is an extension of the traditional graphical method to incorporate multiple hot streams. In this method, the system's cumulative enthalpy/temperature curves are plotted for the hot and the cold streams, and these are manipulated to maximize the recovery. The method provides a useful picture of the ultimate system heat recovery potential, but it loses the information about the individual heat exchangers given by Figure 6.19. Nor does the method define how the individual stream matches and splits are performed to achieve this maximum recovery or consider optimum approach temperatures (Linnhoff et al).

A second potential for heat recovery is the condenser/boiler situation, where heat loads are virtually identical but the temperature driving forces are a few degrees (5 - 30°C) in the wrong direction.

Heat pumps provide the method to make this heat available. By compressing the vapour leaving the top of the column, its condensing temperature can be raised above that required by the boiler. The boiler can then be used to condense the vapour and thereby be supplied with sufficient heat for its vaporisation duty (see Figure 6.20).

The capital cost of such a system is considerable. A compressor is needed, and possibly a standby compressor, and probably a much greater total heat transfer area, because although one unit is saved, the temperature driving forces become very small. The energy required is a fraction of the total heat load that would have to be supplied with no heat pump, and is usually in the form of electrical energy to drive the compressor.

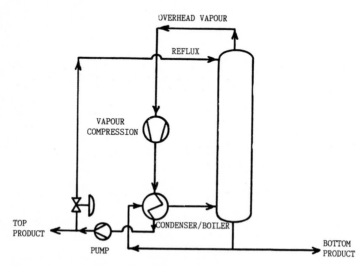

FIGURE 6.20
INSTALLATION OF A HEAT PUMP TO RE-USE CONDENSER HEAT

The system is much discussed but rarely applied. The capital costs are high, and, although the energy saving is high, as electricity costs more than steam, the actual running cost savings are not so attractive. As electricity is pro- duced from steam with only a 50% efficiency, the whole concept is not very sound if the electricity and steam originate from the same fuel source.

The coefficient of performance (COP) of the heat pump - i.e. the ratio of heat pumped to energy required to drive the compressor - can be considerably im- proved by reducing the temperature difference over which the heat has to be pumped. This can be done for part of the heat load if the vapour is taken out of one part of the column and used to vaporise liquid a little lower. This suggestion results in a scheme with excellent COP, part of the total heat load being replaced by the pump. This proposal is made possible because often the vapour flow is critical in a relatively small section of the column - for exam- ple, by the 'pinch' across the feed. If the heat pump works over this section, temperature differences are small. Over the rest of the column, reduced vapour flows (excluding the fraction supplied by the heat pump) are adequate for the separation - save possibly for the inclusion for a few more plates.

This system is easily calculable, simply by assuming that the heat pump is a combination of vapour sidestream and vapour feed. Again the system is rarely applied. Possibly the investment cost of the heat pump to replace a fraction of the total energy - be it very efficient - is not economic.

The principle of supplying vapour only over the critical region of the column has a useful implication for heat recovery systems where hot stream temperature levels do not quite reach the necessary boiler temperature levels. When heat can be recovered, except for a temperature a few degrees too low, an auxillary boiler can be supplied in the column part-way up where the tempera- ture levels are suitable for using the heat source for vaporisation. Liquid is removed from the column at the appropriate point and returned to the same point as vapour.

Such a system is easy to simulate with sidestream and feed studies, and the effect on the total separation can be determined. If this second boiler is placed below the critical part of the column (where temperature profiles are nearly vertical), the effect of supplying, say, 50% of the vapour from this point is often quite acceptable. Hence, a heat recovery scheme can be used al- though the temperature levels at first glance are not compatible.

6.4b MULTIPLE COLUMNS

In theory the potential exists in multiple column systems to use a cascade ap- proach, as in multiple effect evaporators whereby the condenser from one column heats the boiler for the next, and so on down the column sequence. There are practical problems:

- Are the heat loads appropriate?

- Are the temperature levels suitable?

- Do control and stability problems occur?

- Is the resulting system flexible enough in its operation?

Finding a theoretical solution to such a problem is a horrendous task. There is a column sequence problem, overlaid with a heat integration problem, together with the manipulation of column pressure as a further variable to obtain suit- able temperature levels for heat recovery.

In practice, the problem must be solved in stages;

(a) Select a suitable sequence;

(b) list all hot and cold streams, heat loads, and temperature levels;

(c) propose appropriate matches;

(d) identify what process modifications (e.g. pressure or sequence) might produce more attractive matches;

(e) evaluate attractive schemes considering their implications for control and
flexibility and select the best;

(f) try to evaluate a better scheme by (d) and recycle until your time and pa-
tience are exhausted.

From the theoretical standpoint, this is not a very satisfactory process synthesis procedure, but at least it produces a design that could be built and
operated.

6.5 AZEOTROPIC SYSTEMS

Non-ideal systems have liquid activity coefficients that are a function of
concentration. This means that in any system where the activity coefficient
changes are greater than the ratio of the pure saturated vapour pressures of
the key components it is likely that the light at one concentration range will
become the heavy in another. In such a system, it is impossible to achieve complete separation because there must be a concentration where liquid and vapour
have the same composition - and an azeotrope is formed, from which concentration no amount of distillation will induce separation.

For a binary separation, an azeotrope exists when y = x.

Since:

$$y = \frac{P_1}{\pi} = \frac{\gamma_1 P_1 x}{\gamma_1 P_1 x + \gamma_2 P_2 (1-x)} \tag{6.2}$$

at the azeotropic point:

$$y = x \tag{6.3}$$

therefore:

$$\gamma_1 P_1 = \gamma_1 P_1 x + \gamma_2 P_2 (1 - x) \tag{6.4}$$

$$\gamma_1 P_1 (1 - x) = \gamma_2 P_2 (1 - x) \tag{6.5}$$

therefore:

$$\frac{P_1}{P_2} = \frac{\gamma_2}{\gamma_1} \tag{6.6}$$

Notice that whether the azeotrope forms or not, and at what concentration is
only a function of the non-ideality and the relative volatility. Near ideal
close boiling mixtures may form an azeotrope, or wide boiling non-ideal systems. It is not a special characteristic of a certain type of system or
chemical interaction; it is simply the outcome of the VLE data.

To separate azeotropes, the VLE has to be modified. This can be done by
modifying the pressure, where relative volatility changes in a different way to
the ratio of activity coefficients, and the azeotrope point shifts. This shift
in azeotropic point then enables the azeotropes to be 'broken' and pure compo-

nents to be obtained by operating two distillations at different pressures.
Figure 6.21 shows such a system.

If one kmol of feed at concentration x_F is fed to column 1 at pressure P_1, producing a pure fraction of the heavy component and a light boiling azeotrope of
composition x_1 , this can be fed to the second column operating at pressure
P_2, at which pressure the azeotrope x_2 forms of a lower concentration than x_1.
Hence this second column produces x_2 at the top and pure component at the
bottom. x_2 is recycled to the first column, where x_1 is produced together
with pure heavy component that again leave the bottom of the first column.

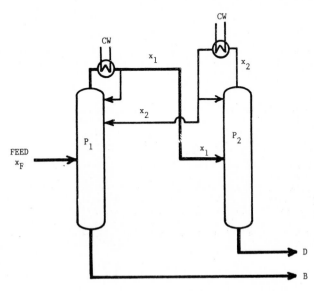

FIGURE 6.21
AZEOTROPES SPLITTING BY WORKING AT TWO PRESSURES

The azeotrope is effectively broken, but the cost is high. If there is a 10%
shift in the azeotropic point as a result of the pressure difference, 10
units of material must be recycled to recover 1 unit of pure product.

Another way of 'breaking' the azeotrope is to add a third component, which so
disturbs the VLE that the azeotrope no longer exists. The general term for
this procedure is extractive distillation, and it can take a number of forms.

The well known method for the dehydration of alcohol is to add a third component (e.g. benzene) to the ethanol/water azeotrope, where the water is so non-
ideal that its activity coefficient lifts it from being a heavy component to a
light component at low concentrations. As the water concentration increases up
the column its activity decreases until it again forms an azeotrope benzene.
Hence the overheads from the system are water/benzene and the bottom product is
pure ethanol. The azeotrope has been broken, but we are left with a
benzene/water azeotrope! Luckily this azeotrope separates into two phases on
condensing and thus this azeotrope can be separated by decantation. A further
column is required to clean the decanted water from benzene and ethanol (see
Figure 6.22).

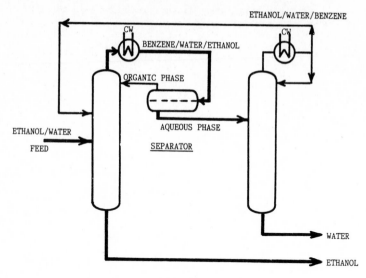

FIGURE 6.22
ETHANOL WATER AZEOTROPE STRIPPING WITH BENZENE

Extractive distillation can also result in a third component forming a high boiling mixture in the base of the column, and the pure components at the top. As long as the column is so operated that the bottom product is not an azeotrope, it can be fed to a second column, where the pure components can be separated and the extractive component recycled to the first column as shown by Figure 6.23. An example of this process is the separation of benzene from cyclohexane by the addition of aniline.

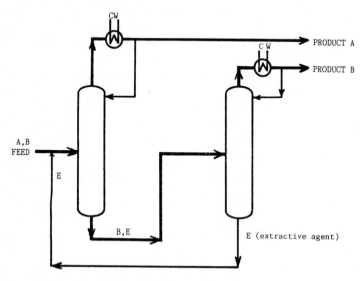

FIGURE 6.23
EXTRACTIVE DISTILLATION WITH HIGH BOILING EXTRACTIVE AGENT

It must be emphasized that these processes are not special forms of distillation, but the result of a perfectly normal separation occurring with the components involved. The normal multicomponent distillation simulation program can be used for the design of such programs in a normal way. The only exception is when two liquid phases exist, and the distillation simulation program available does not have provision for handling two phases and decantation.

Having once gained an impression of the concentration levels attainable by such separations, one can make some algebraic mass balances to determine how the two columns would work together. An ideal course would be the use of a flowsheet program, which would solve the algebraic mass balance problem simultaneously with the distillation simulation.

The trick with extractive distillation is to be able to choose a suitable third component. No procedures exist. The first step is, with much chemical intuition, to locate components that will be sufficiently non-ideal with one of the components to cause required separation. The systems are usually devised experimentally in the laboratory, but now that we have data banks and UNIFAC prediction methods available it is time some systematic method was devised for generating suitable extractive distillation systems.

6.6 SPECIFIC SYSTEMS

Given sufficient time and incentive, distillation systems for specific processes become very complex, as a result of energy and investment savings, or output improvement and relevant only to the separation for which they were devised.

It is salutary to study these systems, to see what was developed years ago from a physical intuition of the system without the use of the computer, and to remember, when we study combinatorial methods for automatic process synthesis that replace human understanding and creativity, that there is more to do than simply couple an optimisation procedure to a distillation calculation.

6.6a PIPESTILLS

The pipestill separates crude oil into the various petroleum fractions, which after further treatment and blending, are distributed as the various petroleum products. Many hundreds of pipestills have been built, and they handle by far the greatest tonnage of all distillation processes. The slightest energy improvement represents enormous savings, company wide. The pipestill has probably had more attention then the other distillation column.

Figure 6.24 is a line diagram of a pipestill. It has the following features that are the result of years of development:

- highly refined feed pre-heat to maximum heat recovery

- extensive use of side-stream take-off to reduce the number of columns

- use of steam stripping to avoid high temperatures at the column base and for light stripping of each side-stream

- use of intermediate pump-around to adjust L/V ratios in the different column sections to optimize the product quality.

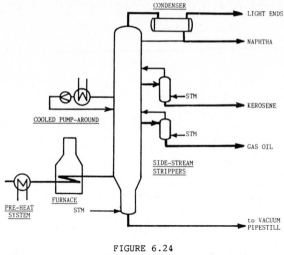

FIGURE 6.24
THE PIPESTILL

6.6b BIOSPIRIT - ETHANOL BY FERMENTATION

In the 1940s, numerous plants were operating to produce ethanol for transportation from fermentation of carbohydrates. Plants were installed to produce
ethanol from wood and ethanol from waste products from paper manufacturing
plants, and even in 1984 some of these ethanol plants were still operating.

The economics of the whole process depend on recovering the ethanol from its 5
- 6% solution after the fermentation. The resulting distillation plant looks
quite complex, as Figure 6.25 shows. The main product is ethanol, but there are
others: acetaldehyde as lights and higher alcohols (fusel oils) as heavies. The
liquor from the fermentation contain disolved carbohydrates, lignin and other
biological matter after centrifuging out the yeast.

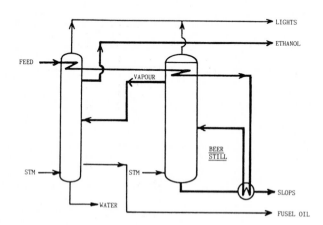

FIGURE 6.25
PRODUCTION OF ETHANOL FROM DILUTE SOLUTION AFTER FERMENTATION

The distillation plant has the following interesting design features:

- live steam is used for column heating;

- a short pre-concentration column is used so that the greater part of the installed plant works with a smaller volume of clean distillate;

- the product is taken from a sidestream to enable the lighter acetaldehyde to concentrate as the distillate and then recycled to the fermentation;

- the fusel oils are also recovered by side-streams;

- extensive heat integration; in particular the dilute cold feed stream is used as the main column condenser coolant and also the effluent stream coolant before it is fed to the column.

These are only some of the refinements developed for these biospirit plants designed in the 1930s. It would complicate Figure 6.25 unnecessarily to include the vacuum flashing of the slops using the live steam feed to the beer still in a steam ejector to both produce the vacuum and recover the flashed steam. Also the partial condensation part way up the column to improve temperature levels for the feed pre-heat, and the number of plates between the top product take-off and the lights removal, all show evidence of considerable initiative and deep study.

It is difficult to believe that any automated synthesis procedure will be developed that will introduce such fine points into a design. The engineer will always have a role to play.

6.6c LIQUID AIR

The production of liquid oxygen and nitrogen by distillation of air is a special process in that it is at low temperatures: heat losses must be minimized, and refrigerated cooling is expensive.

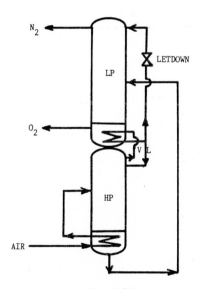

FIGURE 6.26
FRACTIONATION OF LIQUID AIR

The result is that very compact, integrated distillation equipment has been developed, as shown in Figure 6.26.

The following features are of note:

- pre-cooling of the feed air by using it to vaporize the boiler contents of the first column;

- the compact use of the condenser/boiler between columns 1 and 2; the column pressures are so adjusted that the required temperature levels are attained;

- the separation is divided into 2 columns because this enables liquid expansion to provide refrigerant at required temperature levels to replace external refigeration equipment;

- special small-hold-up closely packed distillation trays have been developed to keep column height (and hence surface area and heat losses) to a minimum.

6.7   FURTHER READING

Billet, R., 1973. Industrielle Destillation, Verlag, Chemie, Weinheim.
                - considerable discussion of distillation systems used in practice.

Freshwater, D.C. and Hendry, B.D., 1975. Chem. Engineer, London, p.533.
                - investigation of some column sequencing rules.

Henley, E.J. and Seader, J.D., 1981. Equilibrium-Stage Separation Operations in
                            Chemical Engineering, Wiley.
                - for the synthesis of distillation sequences.

Holland, C.D., 1981. Fundamentals of Multicomponent Distillation, McGraw-Hill.
                - for the consideration of complex columns with sidestreams.

Linnhoff, B. et al, 1982. Users Guide on Process Integration for Efficient Use
                        of Energy, I.Chem.E., London.
                - heat recovery systems in general.

6.8   QUESTIONS TO CHAPTER 6

(1) What different types of boiler are possible, and what are their advantages and disadvantages?

(2) Where should the condenser be mounted?

(3) What are the various criteria for sizing tankage and intermediate storage?

(4) When can the use of sidestreams be effective?

(5) What special equipment is available for vacuum distillation and why is such equipment necessary?

(6) Why is column sequencing important, and what methods are available to determine the optimum sequence?

(7) What methods can be employed to separate azeotropic mixtures into pure components by distillation?

Chapter 7

COLUMN INTERNALS

So far we have concerned ourselves with calculating the number of theoretical plates to achieve a required separation. This is so strongly emphasized that the fact that the distillation is fundamentally a mass transfer process is almost forgotten. The number of theoretical stages is the number of times mass transfer equilibrium has to be achieved between the gas and the liquid streams to achieve the separation. The distillation equipment is the mass transfer equipment required to produce the degree of mass transfer equivalent to the required number of theoretical stages. This is achieved in plate column by bubbling the vapour through a pool of liquid a number of times. Packed columns achieve this mass transfer by counter current flow of the liquid and vapour streams, providing adequate surface area for the mass transfer.

Column internals are therefore to enable mass transfer to occur so that the required number of moles transfer from liquid to vapour and reverse within the contact time available. This demands certain liquid and vapour flows, and the internals must be capable of handling these flows, still maintaining the mass transfer function - i.e. still maintaining the correct hydrodynamic flow regime. The flow of vapour requires there to be a pressure drop across the column internals. For a difficult separation requiring many theoretical stages, this pressure drop can be considerable, and this leads to an increase in the boiler temperature and often to a reduced separation, because it is usual for relative volatilities to decrease at high temperatures. In many chemical engineering processes thermal instability at high temperatures limits maximum boiler temperatures and makes it essential that distillations be carried out under vacuum. For such systems, the pressure drop across the column internals is of utmost significance in determining the operating conditions of the column.

Hence there are three factors that must be calculated for all column internals before a choice can be made and a design can to be specified:

- the mass transfer efficiencies defined as
        "plate efficiency" for plates or
        HETP (height equivalent to a theoretical plate) for packings,

- the maximum vapour loadings at the appropriate V/L ratio,

- the pressure drop at the design conditions.

7.1 MASS TRANSFER THEORY

The two-film theory of mass transfer between phases assumes that there are two stagnant films at each side of the interface and that the component with bulk gas concentration y diffuses by molecular diffusion to the surface and from there across the liquid film to the bulk liquid concentration x. This builds up interface compositions y* and x*, which are assumed to be in equilibrium with each other;

$$y^* = m \ x^*$$ (7.1)

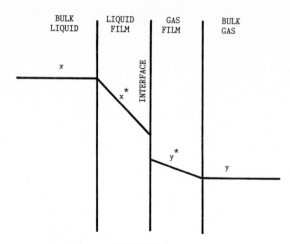

FIGURE 7.1
MASS TRANSFER ACROSS TWO FILMS

With the single film mass transfer coefficients being k_g, k_L respectively, over an interfaced area a, the moles of component transferred per hour (n) are given by:

$$n = k_g a(y-y^*) = k_L a(x-x^*) = K_g a(y-mx) = Ldx = Vdy \quad (7.2)$$

where K_g is the "overall mass transfer coefficient" $kmol/m^2 \, h$. Combining Equation (7.2) and (7.1) to eliminate x* and y*, we get

$$\frac{1}{K_g} = \frac{1}{k_g} + \frac{V\,m}{Lk_L} \quad (7.3)$$

Hence, the mass transfer rate across the interface depends on the mass transfer across each of the films and the slope of the equilibrium line m.

The mass transfer across each film is a result of molecular diffusion. This is measured by the molecular diffusivity (D) of the component y in the particular solvent at the particular conditions. The relation between the mass transfer coefficient and the molecular movement as defined by the molecular diffusivity is dependent on the condition of the "film": the film thickness and its age.

If the film stays permanently at the surface, the relationship is simply to change units:

$$k = \frac{D}{t} \quad (7.4)$$

where t is the film thickness and D is the molecular diffusivity.

If the surface is not in steady state but is continually being renewed by eddy currents that reach the surface, a steady state is not achieved. Higbie showed by using Fick's law that the mass transfer coefficient is related to the square

root   of the molecular diffusion coefficient and the average age of the surface
(t').

$$k \propto \sqrt{\frac{D}{t'}}$$ (7.5)

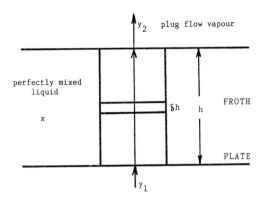

FIGURE 7.2
MODEL FOR MASS TRANSFER ON PLATES

7.1a   MASS TRANSFER ON PLATE

If   we consider a small section of a plate (see Figure 7.2), we can consider it
to be a part of pool of well mixed liquid of concentration x, with a   gas   with
concentration y_1 flowing through it, leaving with a concentration y_2.

The resulting gas/liquid froth on the plates has a height of h (m) with a   surface area for   exchange of, a m^2/m^3 of froth. Assuming the plate section under
consideration has a surface area of a' m^2 and the gas flow rate is V kmol/h m ,
the mass transfer driving force at any point is y - mx and the   mass transfer
over a slice of the froth height ∂h is:

$$a'V \frac{\partial y}{\partial h} = K_g a \, a'(y - mx) \ V \ kmol/h$$ (7.6)

Hence, the transfer over the whole froth given as a gas concentration change is
given by:

$$\int_{y_o}^{y} \frac{dy}{(y - mx)} = \frac{K_g a}{V} \int_{o}^{h} dh = NTU$$ (7.7)

Equation(7.7) defines a number of transfer units (NTU) involved in mass   transfer. It is useful in discussing mass transfer effects, because NTU is a linear
measure of the mass transfer.

Integrating Equation (7.7):

$$- \ln \left(\frac{y - mx}{y_o - mx}\right) = \frac{K_g ah}{V} = NTU \tag{7.8}$$

$$\left(\frac{y - mx}{y_o - mx}\right) = e^{-\frac{K_g ah}{V}} \tag{7.9}$$

Now the point plate efficiency (E_{og}) is defined as the actual concentration change in the gas as it passes through the plate compared with the concentration change that achieves equilibrium:

$$E_{og} = \frac{y_o - y}{y_o - mx} \tag{7.10}$$

$$1 - E_{og} = \frac{y_o - mx - y_o + y}{y_o - mx} = \frac{y - mx}{y_o - mx} \tag{7.11}$$

We can now re-express Equation (7.8) in terms of point plate efficiencies:

$$E_{og} = 1 - e^{-\frac{K_g ah}{V}} \tag{7.12}$$

The object of this derivation is to show the factors that are involved in the plate efficiency achieved by a system. Equations (7.3), (7.5) and (7.12) show that the factors expected to be important in plate efficiency are:

D   the molecular diffusivity in liquid and vapour, which in turn are influenced by fluid physical properties and molecular volumes.

m   the "slope of the equilibrium line" (see Equation (7.1)), which is determined by the fluid physical properties and the molecular volumes.

a   the specific interfacial area/unit volume of froth. This is a function of a plate design. It will also be related to the physical properties of the system because different systems have different froths with different characteristics. "a" will be related to the plate pressure drop in that a higher pressure drop will be required to create more interfacial area.

h   the froth height, which is dependent on the height of the weir that holds the liquid on the plate. It will also be dependent on the liquid flow rate, because the hydrostatic head increases as the liquid flow over the plate increases.

V and L the higher the rates the more mass transfer is required to achieve the same approach to equilibrium. This is to a large extent compensated for by the increased area created at higher liquid and vapour rates.

Equation (7.12) has a further characteristic of importance in discussing plate efficiency: it is extremely non-linear.

The number of transfer units (NTU) as defined in Equation (7.8) is linear, with double the NTU achieved by the plate with double the liquid height.

We can show the effect of increasing linearly the mass transfer performance of the plate and see its effect on the plate efficiency by using Equation (7.12).

Table 7.1 shows the results of such an exercise:

TABLE 7.1

RELATION BETWEEN NTU AND PLATE EFFICIENCY

No. of Transfer Units	Point Plate Efficiency
NTUg	Eog
0.5	0.39
1.0	0.63
1.5	0.78
2.0	0.86
3.0	0.95
4.0	0.98

This non-linearity leads to two particular conclusions:

- It is not worth increasing plate efficiency by increasing froth height beyond a certain point.

- Point plate efficiencies for distillation systems are surprisingly constant, giving a standard plate efficiency of around 0.7 (1.2 NTU).

Hence to increase or decrease the mass transfer by 50% will cause a change in plate efficiency to 0.45 and 0.83 respectively, and not 0.35 to 1.05 as would be the case if the relationship were linear.

Having analysed the factors affecting plate efficiency we understand why the effects are quite small. Distillation plate efficiencies are very similar, because of the similarity of the variables for difficult distillations. Easy distillations require so few plates that the question of accurate plate efficiency is not of interest.

Organic liquid mixtures at their boiling points have similar physical properties and molecular diffusivities. Difficult distillations have L/V ratios near 1.0 and equilibrium line slopes (m) also near 1.0. Columns are designed near flooding, and so all systems have similar values of V. The non-linearity of Equation (7.12) means that there is a fairly sharp point in terms of froth height beyond which the benefits to the efficiency do not justify the increased pressure drop.

All this means that distillation plate efficiencies might be difficult to predict exactly, but they vary over a comparatively narrow range - 0.6 to 0.8 for
example - and once one has a plate efficiency measured for one system then the
prediction of the efficiency from the same equipment with a different system is
not very different: probably within ± 10%.

This is contrary to the general belief, even though the research projects of
Fractional Research Institute (FRI) and earlier the Michigan Tray Efficiency
Research Program, have tried to clarify the matter.

MURPHREE PLATE EFFICIENCIES

So far we have been discussing the plate point efficiency for a small section
of a plate with constant liquid composition. A measure of more practical interest is the effective plate efficiency over the whole plate, and not over a
point on the plate. This is called the Murphree Plate Efficiency and it
is defined similarly to the point efficiency, but with exit and inlet plate
compositions being used in the expression:

$$E_{MV} = \left(\frac{y_{in} - y_{out}}{y_{in} - mx} \right)$$

(7.13)

The major difference between the point and Murphree plate efficiencies are due
to the flow pattern on the plate. The liquid flows across the plate, and
this gives an improved separation compared with a complete back-mixed situation. Theoretically considered cases, assuming zero liquid mixing, show that
enhancement of around 50% can be expected. In practice, however, such enhance-
ments are much smaller because of liquid back-mixing which reduces this effect.

Various correlations have been produced that are based upon predicting the
liquid eddy diffusivity from the plate dimensions and conditions, and this is
used via a Peclet number to determine the enhancement (AIChE Bubble Tray design
method). This effect is noticeable on large trays where efficiencies sometimes
over 100% are quoted. On trays less than 1 m diameter the effect is small. It
is important, however, to appreciate that plate efficiencies improve with larger diameter columns and that methods exist for predicting the effect of column
diameter on plate efficiency.

OTHER FACTORS AFFECTING MURPHREE PLATE EFFICIENCY

So far our plate model allows for cross-flow of liquid over the plate and plug
flow of vapour through the froth layer. When this model is not adhered to, this
results in a reduction in plate efficiency.

At higher vapour rate, liquid droplets are entrained in the vapour stream and
are returned to the plate above. This phenomenon is called liquid entrainment.
It has a completely negative effect on the separation efficiency because the
enriched liquid is simply mixed back again with the leaner liquid. The reduction in plate efficiency is a function of the fraction of liquid entrained, and
beyond a few percent entrainment there is a marked fall in plate efficiency.
This fall in plate efficiency precedes the flooding point of the column where
so much liquid is entrained that less can leave the column than is fed. Hence,
the operating range of the column is less than the physical flooding point because of this reduced plate efficiency.

At low vapour rates with most designs of tray, the liquid cannot be supported on the plates and "weeping", occurs whereby some of the liquid passes through the plates (bypassing vapour/liquid contacting and causing vapour instabilities), which causes a reduction in the overall plate efficiency, giving a minimum useful operating point for a plate.

These effects determine the operating range of a plate. Figure 7.3 shows this effect qualitatively.

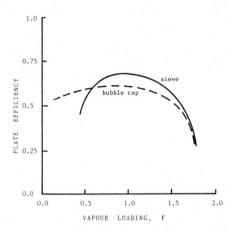

FIGURE 7.3

QUALITATIVE DESCRIPTION OF THE OPERATING RANGE OF A PLATE

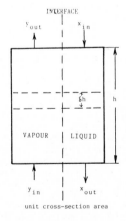

FIGURE 7.4

FLOW MODEL FOR MASS TRANSFER OVER PACKING

As will be discussed later, different plate designs have different weeping and entrainment characteristics and different "flexibility" characteristics with respect to vapour rate changes. The vapour rate weeping and entrainment points are influenced by liquid flow, fluid properties, and plate detailed dimensions (weir heights, hole size etc.). Hence these factors affect the ultimate capacity of a plate column, since this is determined by the entrainment.

7.1b MASS TRANSFER IN PACKING

The flow model for gas and liquid over a packing is shown by Figure 7.4. In this case the liquid is counter-current to vapour and not perfectly mixed or cross-flow as discussed for plates. Taking a slice of column height δh (unit cross-section area), the area available for mass transfer is $a\,\delta h$ (where a is the specific interface per unit volume of packing (m^2/m^3), and the moles transferred per hour δn are given by:

$$\delta n = K_g a\ \delta h(y - mx) \tag{7.14}$$

since the change in mole fraction δy is defined as $\dfrac{\delta n}{V}$, Equation (7.14) can be expressed as:

$$\frac{dy}{(y - mx)} = \frac{K_g adh}{V} \tag{7.15}$$

and the performance of a complete column determined by the integration of Equation (7.15).

$$\int_{y_{in}}^{y_{out}} \frac{dy}{(y - mx)} = \frac{K_g ah}{V} \tag{7.16}$$

This equation is analogous to Equation (7.7) for mass transfer over a plate; the LHS defines a number of transfer units required by the defined separation, and the RHS relates the transfer units required to the column height (h) and packing characteristics ($K_g a$). This equation, however, differs from Equation (7.7) in that the x is not constant but is a function of y, L and V. Hence, the integration of Equation (7.16) is no longer a simple matter and numerical (or graphical) methods must be used to determine a number of transfer units equivalent to the required separation, and this can then be easily related to the height required. This design approach (using NTU) is used for absorber and stripper mass transfer where one or two theoretical plates are involved, but it is more difficult to apply when many plates are involved, as in distillation. Furthermore, the earlier chapters have dealt with determining the number of theoretical plates required for a required separation, and any practical column design method should follow on from this approach and not demand the separation be recalculated in terms of NTU's.

For these reasons, the NTU approach is not use for designing packed distillation columns, but the "HETP" (Height Equivalent to a Theoretical Plate) is determined as the basis for the design. The HETP is the height difference between the points in a packed column where the gas composition is in equilibrium with the liquid composition.

It is not useful to define HETP in terms of NTU's, but generally a theoretical
plate in a counter-current packed column is equal to about two transfer units.
Hence, the HETP of a packing is a function of $K_g a$ for the packing, where a is
given by the dimension of the packing and K_g by the hydrodynamics and fluid
velocities and physical properties. Thus HETP is affected by molecular
diffusivity and liquid and gas film controlling resistances as defined by Equation (7.3) and (7.5). It can also be expected that the HETP is affected by
densities, surface tension and flow rates, but these effects must be determined
experimentally.

The overall effect of the gas flow rate, with its two compensating factors of
improving K_g but needing far more mass transfer to achieve the same concentration changes, is to give an approximately constant HETP over the hydrodynami-
cally acceptable gas flow range of a packing, as is shown by Figure 7.5.

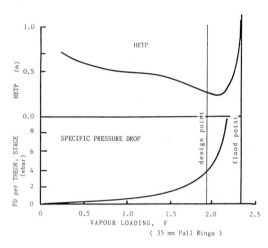

FIGURE 7.5
QUALITATIVE RELATIONSHIP BETWEEN HETP AND VAPOUR VELOCITY

The best separation efficiency is achieved near the maximum hydrodynamic flow
capacity of the column, just before the effect of increased entrainment reduces
the overall separation efficiency. For practical purposes, the operatimg point
is generally chosen to be 80% of the flood point. This gives an operable
plant and also benefits in terms of pressure drop required per theoretical
stage.

Whereas prediction methods exist for plate efficiencies, the prediction of HETP
for packing separations are by no means so well developed. When a manufacturer
introduces a new packing he quotes measured HETP values obtained using a test
distillation system, and this value is used for whatever system is to be
distilled.

In practice, the same comments apply to the similarity of HETP's for various
systems as to the similarity of plate efficiencies. Physical properties and
molecular diffusivities are similar for most distilling systems. L/V and m are
both near unity when the distillation requires many stages; hence, the transfer
of the HETP measured from one system to another is not unjustified.

HETP is influenced by column diameter and packed height in that these factors affect the distribution of liquid and vapour and hence the flow model describing the system. Maldistribution occurs when the packing is large compared to the column because the liquid has not the opportunity to divide a sufficient number of times to achieve an even distribution. A column/packing diameter ratio of higher then 10:1 is preferred to ensure that maldistribution from this source does not occur. Large depths of packing also give rise to poor liquid distribution, in that there is a tendency for the liquid to migrate to the walls and to give a freer path to vapour in the column centre. Measures against this are to put wall rings in the column to divert the liquid from the walls or to completely redistribute the liquid at every 2 - 3 column diameters.

If measures against liquid maldistribution are not taken, a higher HETP than that given by the packing manufacturer is to be expected.

Packing can also suffer from inadequate 'wetting' when very low liquid rates are involved. At low liquid rates only a part of the packing surface area is wetted, and this reduces the mass transfer performance of the packing. This effect is important for absorption/desorption where liquid rates can be very low, but for distillation where the liquid rate has a close relation to the gas loading, there are no reports of low HETP values due to poor wetting. (Although it could be suspected that for incompatable systems such as polypropylene rings with aqueous systems, poor HETP values could result). The fact that in distillation the system is at its dew point may also contribute to the apparent unimportance of wetting rate in distillation.

7.1c GENERAL COMMENTS ON THE PREDICTION OF SEPARATION PERFORMANCE

There is a widespread belief that plate efficiencies vary over very wide limits (0.2 - 1.2), are completely unpredictable, and therefore must be determined for each system experimentally.

This attitude, which is based on an incomplete assessment of the situation has a devastating effect on confidence in distillation design: if the plate efficiency can have six-fold variations, then experiments are necessary; if experiments are necessary, then little is to be gained from theoretical computations, particularly the more complicated studies; hence, distillation should remain a purely experimental and empirical subject, except for the oil refining and petroleum industry.

If plate efficiencies and HETP can be predicted reasonably accurately, the way is open for a more rational design approach to distillation in the general chemical industry.

The misconception that distillation plate efficiencies are unpredictable is based upon the following points:

- An experimentally measured "plate efficiency" contains all the errors of prediction of VLE. If "ideality" is assumed, or low concentrations where the VLE has never been measured are used, the resulting "plate efficiency" may be a useful parameter for designing the same system, but it is not E_{mv} or E_{ml}.

- If a system involving the absorption or stripping of very soluble or slightly soluble components is used to measure the plate efficiency, wide variations can be expected because the mV/L term is not near unity, and this affects the efficiency, not on mass transfer grounds, but simply on mass balance and or unit conversion grounds. For example, a desorption of oxygen may have 0.9 plate efficiency (based on liquid composition approaches - E_{ml}) but when converted to be expressed in vapour composition terms (E_{mv}) a figure of 0.2 is obtained.

- Any malfunctioning of equipment shows itself in the 'plate efficiency';
        - maldistribution, entrainment;
        - ill-conceived testing equipment with additional mass transfer areas;
        - inaccurate flow measurement.

If one confines a plate efficiency survey to distillate systems, published figures are much more consistent, and values between 0.6 and 0.9 are most usual (as shown in Table 7.2), with higher efficiencies being associated with column diameters over 1 m.

TABLE 7.2

PLATE EFFICIENCY FIGURES FOR DISTILLATION SYSTEMS REPORTED IN THE LITERATURE

PLATE TYPE	SYSTEM	DIA (m)	MEASURED EFFICIENCY (Emv)
Bubble cap	ethanol/water	0.75	83 – 87
Bubble cap	methanol/water	1.0	90 – 95
Bubble cap	benzene/toluene	0.5	70 – 80
Sieve	ethanol/water	0.75	75 – 85
Sieve	methanol/water	1.0	90 – 100
Sieve	benzene/toluene	0.5	60 – 80
Valve	ethanol/water	0.75	75 – 83
Valve	methanol/water	1.0	70 – 80
Valve	benzene/toluene	0.5	75 – 80
Bubble cap	ethylbenzene/styrene		67
Sieve	"		70
Valve	"		76

Based on this evidence, the error in assuming that all plate efficiencies are 0.75 involves a maximum error of ±20% – a very different picture from the 6-fold variation often quoted.

If we make appropriate allowance for plate diameter, changes in system (D, mL/V) and plate type, it is reasonable to expect predictions of plate efficiency of ±10% be achievable, particularly if a measured point, with any system, is available as a reference point. This accuracy is no worse than any of the other data for the distillation design.

As previously pointed out, the relative constancy of the efficiency distillation is because mV/L is near unity, properties of distillation systems are similar, and the form of the definition of E_{mv} makes it insensitive to mass transfer changes.

Basically, the same argument can be used to explain why given a packing has a relative constant value of HETP for all distillation systems. The major factor affecting HETP variation is the maldistribution: reliable values for HETP for comparison purposes can only be obtained when the liquid is frequently redistributed over the packing.

7.2 SELECTION OF COLUMN INTERNALS

Given that a separation needs a certain number of theoretical plates and a defined reflux ratio, the designer must decide upon a column diameter and plate
or packing details to achieve the required separation and capacity.

The first step is to choose appropriate column internals. In the first instance
should it be plates or packing? Then, more precisely, which type of plate or
packing?

Much is written comparing the advantages and disadvantages of plates vs packing. These are summarised in Table 7.3. There are many instances where both are
equally suitable, and the choice then depends on tradition within the firm concerned. Chemical companies usually use packing, and it is up to the designer to
explain why he does not want to use packing if he uses plates. In the oil
processing and petrochemical industry, the reverse is true, and reasons for not
using plates must be given. This attitude is justifiable in terms of spares and
company knowhow and should not be attributed entirely to conservatism.

The book of Billet shows in considerable detail the results of comparative cost
studies for different types of column internals for a given duty. Results from
20 different internals showed that the difference in cost of column plus
internals varies by a factor of three, depending on the internals chosen. Of

TABLE 7.3

COMPARISON OF ADVANTAGES OF PLATE AND PACKED COLUMNS

Advantages of plate column

(1) lower capital costs at large diameters (over approximately 1 m)

(2) less sensitivity to blockage or dirty feeds and easy to clean

(3) preferred to packing at lower liquid rates, as there are no
    liquid wetting problems

(4) preferred to packing at high liquid rates, because vapour
    throughput not so dependent on liquid flow.

(5) heat removal possible by incorporating cooling coils on plates.

Advantages of packed column

(1) lower capital costs at small diameter (less than 1 m)

(2) low pressure drop per theoretical stage (important for vacuum
    distillation and thermally sensitive boiler compositions)

(3) more corrosion resistant, with a range of materials of construction
    available

(4) preferred for foaming systems

(5) the low hold-up is advantageous for batch distillation

---

particular interest was that the order of preference changed with the constraints given for the design. The change from carbon steel to stainless steel changed the order of preference, as did the constraint to design to a particular pressure drop rather than a given percent flooding.

The study showed Pall Rings and valve trays to be above average, and bubble caps generally poor. As the allowable pressure drop fell, valve trays lost position, and structured packing economics improved. Billet emphasises that each system at each set of conditions requires its own study to enable a correct selection to be made.

7.3 PLATE INTERNALS

A plate consists of a device by which a liquid layer over the column cross-section is so supported that the vapour can pass through it to enable mass transfer to occur, with the leaving vapour as near as possible at equilibrium with the liquid.

7.3a TRAY TYPES

There are many devices to achieve this, some very thoroughly constructed to ensure that the fluids can flow only in the direction required of them (e.g. the bubble cap) and others very simple, with virtually no provisions to prevent incorrect flow patterns developing (e.g. grid trays). Figure 7.6 gives a selection of such devices.

BUBBLE TRAYS

These are the oldest, most thorough design of plate contactor, with every vapour channel liquid sealed and with a weir to prevent incorrect flow patterns developing. This results in a plate with the widest operating range (with respect to acceptable plate efficiency) and greatest stability with respect to high liquid levels on the plate. All this has to be paid for, both in terms of cost and pressure drop. Bubble cap plates have the highest cost and the highest pressure drop.

SIEVE TRAYS

Sieve tray (Figure 7.6) consists of a plate over which the liquid runs and with holes through which the vapour flows. Under design operating conditions the gas keeps the liquid on a plate and the plate functions well. It has a lower pressure drop than a bubble tray and a slightly better efficiency (the dead volume of the bubble caps can be used for mass transfer).

As conditions deviate from the design - particularly as vapour rates fall - the flow becomes less stable and some liquid falls through the perforations intended for the gas flow. A small amount of "weeping" is not detrimental to the operation, but as this increases, major instabilities can occur, with parts of the tray "dumping" the liquid through the holes, and only part of the tray being active in letting the vapour through.

Sieve trays are preferred to bubble caps in terms of cost, performance, pressure drop and liquid capacity as long as flexibility is not required. Sieve trays are not suitable for columns where a high turn-down ratio is required. Columns with a variation in vapour load over the column are also difficult to fit with sieve trays. Different designed trays are necessary at different levels in the column. Different column diameters may even be necessary. In that case, costs rise and the more stable bubble-tray column may then offer a cheaper solution with a single plate design and diameter over the whole column.

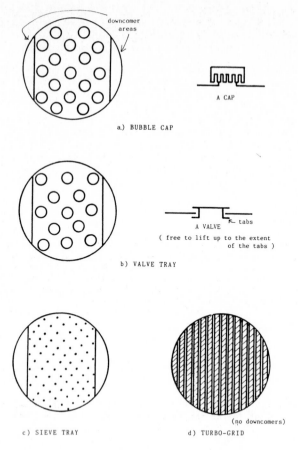

FIGURE 7.6
SOME TYPES OF DISTILLATION TRAY DEVICES

VALVE TRAYS

The valve tray (Figure 7.6) is a development aimed at keeping the flexibility
of the bubble tray but reducing its investment cost. The vapour holes in the
plate are covered by metal flaps (valves), which lift up when vapour passes but
close down at low vapour rates. This regulates the vapour opening and therefore
gives a consistent efficiency and pressure drop over a wide operating range.

Without the full complexity of the bubble cap, the lower operating vapour velocity is not as low as the bubble cap tray.

The movement of the valves has a cleansing action, which frees blockages and
incrustations, and so the valve tray is particularly suitable for dirty duties.

A development of the valve-tray, the sieve-valve tray, is reputed to have improved efficiencies and reduced pressure drop compared with the valve tray it-
self.This tray has in addition to the valves, some perforations as the sieve
tray. Hence, at lower rates it operates as a normal sieve tray, at high rates
the valves open, thus increasing the capacity without increasing the pressure
drop.

As can be anticipated, valve trays have lower pressure drops, lower liquid gradients and lower costs than bubble caps, with high efficiencies, but a lower operating range. Efficiencies and pressure drop are closer to those of sieve plates, but the operating range is a considerable improvement over that of the sieve tray.

TURBO-GRID TRAYS

Turbo-grid trays consist of a plate made up entirely of bars across the whole column with no downcomers equipment provided for liquid flow. The liquid rains down the spaces between the bars and the vapour flows up, thus providing the necessary liquid/vapour contact. As can be expected, these internals are very inexpensive but have a low efficiency and low flexibility for gas throughput changes. Because of the absence of downcomers, all the column area is available for contacting, thus enabling a higher vapour throughput for a given column diameter to be attained.

OTHER TRAY TYPES

Many tray manufacturers provide their own patent designs: Varioflex caps, Perform-trays, film trays, expanded metal trays, sieve-slit trays, tunnel caps and ripple plates, just to mention a few. Each has its own claimed advantages and disadvantages, each are supported by manufacturers performance data. It is not the task of this text to evaluate them, or even to provide comparative data. Much information has been published already (Perry, Billet), but a company carrying out a study would want to contact likely manufacturers directly to obtain their design data and make comparisons for themselves.

7.3b   TRAY OPERATION

Figures 7.7a and 7.7b show the operation of a sieve tray.

The active area in this tray is limited by the geometry and the liquid downcomers to about 70% of the tray area. This is the area in which the perforations (or bubble caps, or valves) are located. At the design conditions, this active area should be operating at optimal efficiency, which occurs at about 80% of the flood point.

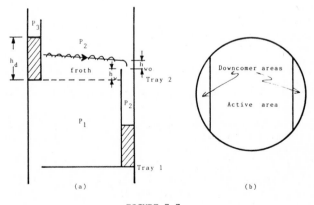

FIGURE 7.7
THE SIEVE TRAY

The liquid flows down the inlet weir from the tray above. In the inlet downcomer the vapour is disengaged, so that only liquid phase flows on to the plate. As the liquid flows across the plate, it is aerated into a froth by the vapour coming through the holes in the tray. At low velocities (0.3 m/s) a stable hexagonal type foam may build-up like suds in a wash-tub. However, at industrial gas velocities (0.6 - 1.4 m/s at normal pressures) such hexagonal foam is not stable and the "froth" on the plates is a highly active mixture of continuous liquid phase with gas bubbling and jetting through it. At still higher vapour loadings the character again changes to a continuous vapour phase jetting through the holes and the liquid in the foam as dispersed droplets finding their way across the plate throughthe jets of vapour. At this point, entrainment is high, and a further increase in vapour rates "blows the liquid off the plate" back to the plate above, and the column floods.

The liquid flows over the outlet weir (whose height determines the minimum froth height on the tray) and then down the downcomer. The inlet weir serves to prevent partial vapour flow up the downcomer, which could completely disturb the plate flow patterns.

Vapour pressure drop is caused by pressure drop through the perforations and through the froth. Liquid pressure drop occurs owing to the liquid flowing down the downcomer and under the downcomer outlet. The stationary height of liquid in the downcomer is then this liquid pressure drop plus the plate vapour pressure drop (by inspection of Figure 7.7). It can happen that at high loadings this total pressure drop is greater than the distance between the plates; the downcomer becomes completely full and still an adequate liquid flow is not achieved. The column is flooded because of downcomer flooding; this is different from the entrainment flooding described previously, which was caused by high vapour rates.

7.3c DESIGN CORRELATIONS

Firstly a column diameter has to be determined. The maximum vapour capacity of most plates lies in the F-Factor range from 1.9 to 2.8 $(kg^{1/2}m^{-1/2}s^{-1})$. Propriatory and company in-house information can be used to give a more accurate estimate of the maximum capacity of specific plate types, with correlations involving ρ_L, ρ_V, σ and V/L.

Secondly a hole size and percent free area has to be determined. For a sieve tray, the holes are selected to have low pressure drop and yet still have no serious weeping problems. The prediction of weeping is based upon a balance of the pressure (or height) of liquid supported on the plate, and the pressure required for bubble formation and vapour flow through the holes.

Based on empirical relationships reported in Perry; to prevent serious weeping:

$$(h_{or} + h_{ba}) \geq 5.0 + 1.8\ f_t(h_w + h_{wo}) \tag{7.17}$$

where, h_{or} (mm liquid)   is   pressure drop of vapour through orifice,
         h_{bu} (mm liquid)   is   pressure to create bubble at orifice,
         h_w (mm)   is   weir height,
         h_{wo} (mm)   is   head of liquid over weir,
         f_f (-)   is   hole area as a fraction of column
                                     cross-section area.

The pressure drop of vapour through the orifice can be calculated using an orifice equation:

$$h_{or} = \frac{1000\ v^2(1 - f_t^2)}{1.62\ g} \tag{7.18}$$

where   v is the vapour velocity through the orifice (m s^{-1}),
and   g is the gravitational constant   (9.81 m s^{-2}).

The pressure to form the bubble is given by

$$h_{bu} = 4.13 \times 10^5 \frac{\sigma}{\rho_L d_h} \qquad (7.19)$$

where, σ is the surface tension (N m^{-1}),
d_h is the orifice diameter (mm),
and ρ_L is the liquid density (kg m^{-3}).

The head of liquid over the weir can be calculated from the standard Francis Weir Formula for a rectangular weir, which can be approximated by;

$$V' = 0.40 \left(\frac{h_{ow}}{1000}\right)^{1.5} L' \sqrt{2g} \qquad (7.20)$$

where, V' is flow rate (m^3 s^{-1}),
h_{ow} is height of liquid over weir (mm),
L' is length of weir (m),
and g is the gravitational constant (9.81 m s^{-2}).

Equations (7.18) to (7.20) enable expression (7.17) to be evaluated to confirm that weeping is not a problem.

The pressure drop of the plate must then be calculated. The total plate vapour pressure drop is the sum of the pressure drop through the dry plate(h_{or}), the pressure to form the bubbles (h_{bu}) and the pressure drop through the liquid on the tray (h_w + h_{wo}). This last pressure drop however, has to be corrected for the froth density being less than the liquid density.

Froth density can be correlated against F-Factor, and is of the order of $0.21\rho_L$ at an F of 1.8; $0.25\rho_L$ at an F of 1.4: and $0.35\rho_L$ at an F of 0.9.

In large diameter columns then allowance should also be made for the hydraulic gradient required for the liquid to flow across the plate. The literature contain correlations which can be used for detailed work. Figure 7.7 is a summary of the pressures and levels involved.

Having calculated the pressure drop and knowing the tray hydraulics, the liquid downcomer flow pressure drop can be calculated using normal pressure drop calculation methods, to enable the height of liquid in the downcomer to be calculated, to check that downcomer flooding will not be encountered.

Calculation of the plate efficiency for a particular design is possible by using correlations developed in 1958 and published as the AIChE Bubble tray design manual. In principle, this method uses the mass transfer theory developed at the beginning of this Chapter, based on molecular diffusion through two films, together with experimental data to provide the mass transfer coefficients.

The height of froth on the plate is calculated from plate hydraulics and this is a major factor in the prediction of plate efficiency.

Corrections are made for enhancement due to cross-flow, more corrections for backmixing, and still further corrections should be made for entrainment before a final Murphree Overall Efficiency is produced.

More details of the design correlations will not be given here. A large company should have them available as a computer program. A medium company should be participating in the FRI research program which has published its own methods. Otherwise the 1958 methods of Gerster and the AIChE Bubble Tray design manual could be used. The information in Perry is based on extracts from this method which can be used if the original literature is not available, and more details are given by Holland.

The EURECHA program INTERN, which is detailed in Appendix 4 is a teaching program for the design of internals, and this allows for the design of specified plates to be checked using the methods described above.

7.3d DESIGN PRINCIPLES

The designer has to specify the design of the plates to be installed (their diameter, perforations or bubble cap dimensions and number), the downcomer design (area, weir size, tray spacing), and the number of trays. The resulting equipment should not flood owing to entrainment or downcomer capacity at the design capacity and should have a turn-down flexibility meeting that specified for the design.

The column must therefore be checked at a number of points in the column (bottom, top, just below and just above feed) to ensure that performance of the trays, both at highest and lowest rates, does not restrict the column design capacity or turn-down ratio.

When this brings problems to light, more than one tray design will be needed in the column. In many cases, this can be achieved by using the same basic design for all trays but inserting "blocking strips" over selected holes by the liquid inlet or outlet to improve the stability in those parts of the column where the vapour velocity is lower. When this does not suffice, either a more flexible tray type must be chosen, or the column diameter must be changed and an expensive conical reduction piece inserted in the column.

The design of trays for a distillation column is an excellent "design exercise" in its own right, requiring calculation, engineering judgement, operational flexibility studies and compromise solutions.

The following steps are needed for the complete design of a single tray:

(1) On the basis of liquid loading, determine whether cross-flow or dual liquid flow systems are required.

(2) Choose a suitable working loading F-Factor ($u\sqrt{\rho_g}$) representing maximum plate efficiency (approximately 80% of the flood point). Alternatively, various correlations exist to determining the flooding vapour capacity from ρ_L, ρ_g, μ_L, and L/G. One such correlation is the Souders-Brown correlation, which determines the entrainment flooding velocity for bubble and sieve trays, in terms of liquid and vapour densities, the L/V ratio, and plate spacing. This particular correlation is used in the INTERN computer program (Appendix 4); it is detailed in Perry.

Determine the column diameter from this design velocity and the required vapour flow.

(3) Choose a suitable outlet weir (e.g. 30 mm) and % active area (e.g. 70%), % hole area of active area (e.g. 8%), and hole size (e.g. 10 mm). Calculate weeping, entrainment and froth height.

(4) Calculate pressure drop and plate efficiency and check weeping. Return to (3) if the resulting plate indicates problems. Iterate between (3) and (4) until a good plate design is obtained. This calculation is best carried out by computer.

(5) Choose downcomer area, inlet gap and weir length, calculate liquid pressure drop and liquid level in downcomer. Repeat until a satisfactory design results with 400 mm tray spacing.

(6) Determine the total number of trays required by calculating the plate efficiency of the proposed plate design. Calculate the tray pressure drop and

resulting total pressure drop. Is this pressure drop acceptable or should we return to (3) to find a design with reduced pressure drop?

(7) Check this design at conditions expected at the other points in the column where major flows and condition changes are expected. Revise (2) and (3) if necessary to enable one design to cover the whole column.

(8) Test the design for turn-down flexibility at the extreme points in the column. If turn-down problems are evident, can a return to point (3) result in a more flexible design? Can a reduction of active area suffice? Should alternative column diameters or multiple diameters be considered?

7.4 PACKING INTERNALS

Packing is a material of high surface area per unit volume and high voidage that is used to fill the distillation column. Liquid runs down the surface, 'irrigating' the packing. The vapour flows through the voids, and mass transfer occurs at the liquid interface.

The liquid should wet the whole area of the packing, which may be difficult to achieve if materials are not compatible (e.g. aqueous systems with polypropylene packing) or if the liquid rates are very low (e.g. well below "mimimum wetting rates").

The effectiveness of the packing is dependent on the specific surface area a (the m^2 of area per m^3 of packing) and the mass transfer coefficient K_g.

A packing with high a has small dimensions and this creates a high pressure drop and a low flooding velocity. Larger size packing with a lower a have higher HETP values leading to taller, narrower columns. The packing size therefore determines the L/D ratio of the column. An L/D ratio resulting in low costs, consistent with good liquid distribution, should be specified for the design. An upper limit to the packing size is given by distribution considerations. When the packing diameter is more than 1/10th of the column diameter, liquid maldistribution is evident. This therefore defines the upper limit to the packing diameter.

The operating point of a packing for normal distillation purposes should be at its best HETP condition, which is about 80% of the flooding point. If vacuum distillation or distillation requiring minimum pressure drop is under consideration, the operating point choosen should be the point showing minimum pressure drop per theoretical plate.

7.4a PACKING TYPES

As with trays, there is a wide variety of packing types available. The more important of these are shown in Figure 7.8, and they are discussed separately in the following sections.

RASCHIG RING

The oldest form of distillation packing material is the "Raschig ring". This is a stoneware ring the height of which is equal to its diameter, and available in basically 8 mm, 25 mm, 35 mm and 50 mm sizes. It is corrosion resistant, cheap, and readily available. Its pressure drop/theoretical plate characteristics are better than those of plate internals, but it is one of the least satifactory amongst the column packing types. This characteristic can be improved if mild steel is used in place of stoneware, because of the thinner wall and resulting better voidage.

a) Raschig Ring   b) Pall Ring   c) Interlox
                                                      Saddle

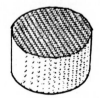

d) Sulzer Structured Packing

FIGURE 7.8
COLUMN PACKING TYPES

PALL RINGS

Pall rings are basically Raschig rings with tabs punched into their side wall. This reduces the resistance to vapour flow of those rings broadside on to the gas flow and causes more turbulence of liquid and vapour, which increase the mass transfer coefficient. This results in a packing that has a lower pressure drop/theoretical plate, higher vapour capacity and slightly improved HETP over the normal Raschig ring. There are further improvements if mild steel Pall rings can be used, because the thinner walls give even better pressure drop characteristics.

SADDLES (Interlox, Berl and Novalox)

These stoneware saddles were developed as an improvement to Raschig rings, in pressure drop, HETP and maximum loading. They have better characteristics but are more expensive. Whether they produce a better design than Pall rings can only be determined by a detailed case study.

SULZER-STRUCTURED PACKING

Sheet material, either metal, plastic, ceramic or woven wire or plastic filament, which is formed into a crimped packing (see Figure 7.8), can be used as a column packing where very low pressure drops are required. This packing has a regular structure, as apposed to the randomness of dumped packing, and it is supplied in sections that fit snugly, directly into the column. It is considered that the regularity of the packing results in its improved characteristics over random packing. The packing has a very low pressure drop and low HETP, resulting in an outstandingly low pressure drop per theoretical plate of about 1 mbar. Though this packing is expensive and sensitive to fouling, it finds application in vacuum distillation and distillation requiring many plates but with restriction of pressure drop (e.g. heavy water distillation, where high pressure drop reduces the relative volatility).

The low HETP and high loadings results in short, small-diameter columns, and if the column has to be made of an expensive material the overall cost of column and packing can be lower with these higher performance packings.

The feeds to columns with these packings have to be filtered.

MATERIALS OF CONSTRUCTION OF PACKING

Original packing material was a stoneware (ceramic). This had the disadvantages
of being rather thick - leading to lower voidages and high pressure drops, and
rather fragile. The fragility means that care must be exercised in filling the
columns and that there is a continual breakage in the column owing to local
flooding and packing movements resulting from control upsets and swings, etc.

Usually, after 1 - 2 years the column has to be dismantled, emptied, and the
stoneware packing replaced. Metal packing has the advantage over stoneware that
it has a lower wall thickness, leading to lower pressure drops, and it is not
fragile. However, it is more expensive, and subject to corrosion.
Stainless steel, or where corrosion allows, mild steel Pall rings have good
properties. Polypropylene rings are offered; they are inexpensive, light and
corrosion resistant. Column emptying is very simple, as the rings can be
floated out. However, they present wetting problems with some systems and have
a lower voidage than steel rings.

7.4b DESIGN PRINCIPLES

The designer has to specify which packing should be used: what size, and material of construction. The column diameter and packed height must also be defined, as must the arrangements for liquid distribution and redistribution.

The choice of packing type is determined firstly by the column diameter.
For good distribution the ratio column diameter:ring diameter (or packing
characteristic dimensions) should be greater than 10:1.

An approximate column diameter can be obtained by looking at the vapour loading
- in terms of loading factor F $kg^{1/2}m^{-1/2}s^{-1}$. This loading factor is of
the order of 2.5 - 3.0 for a fully loaded column with stoneware rings. The
value is dependent upon many factors (L/G, ρ , packing type, etc.), but this
figure gives an initial estimate for the column diameter and hence packing
size.

The choice of packing type depends upon the duty required. A simple,
straightforward distillation could employ Raschig rings, since these are the
cheapest. If pressure drop is a consideration, Pall rings would be preferred to
normal Raschig Rings. If the number of plates is high, a smaller size than
10:1 column diameter would reduce overall column height. If the number of
plates induces pressure drop problems, Pall rings would be a sensible choice.
Metal rings have a lower pressure drop than stoneware, small metal Pall rings
would be a further improvement if corrosion is not a problem.

The Intalox saddles are reported to have good characteristics, thus in addition
to Pall rings, the saddles could be investigated to see if they produce any
cost advantage.

If pressure drop, even with Pall rings, is still a major problem, consideration
should be given to Sulzer-type packing, which has earned a reputation for difficult separations where pressure drop problems occur.

These comments are useful guidelines, but the final selection comes as a result
of considering a number of the possible good proposals in more detail - as far
as costing if necessary.

COLUMN DIAMETER

The flooding point of a packed column is a function of the physical properties
of the system, ρ_L, ρ_g, μ_L, the liquid/gas ratio L'/G, and the packing. The packing is characterized by a 'Packing factor', F_p , which is related to the
geometry of the packing.

$$F_p = \frac{a}{\varepsilon^3} \quad m^{-1}$$

(7.21)

where a is the surface area per unit volume of packing and ε is the voidage per fraction free space in the packed bed. Table 7.4 gives some values of F_p for various packings.

The relationship between the flooding gas velocity and these factors is presented in the form of an empirical, experimental correlation on Figure 7.9.

This is an old method proposed by Sherwood in 1938, which basically remains the standard procedure today. The later improvements by Eckert in 1970 enables gas velocities at constant pressure drop as well as at the flood point to be determined.

TABLE 7.4
PACKING FACTORS FOR VARIOUS PACKINGS

PACKING	NOMINAL SIZE mm	PACKING FACTOR F_p (a/ε) m^{-1}
Raschig Rings (ceramic)	12.5	1740
" " "	25	460
" " "	50	200
Pall Rings (metal)	25	150
" "	50	60
Intalox Saddles	25	300
" "	50	120

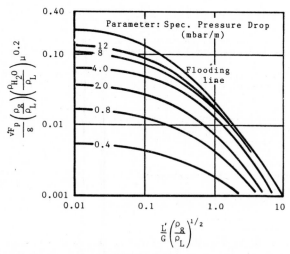

FIGURE 7.9
GENERALIZED FLOODING CORRELATIONS FOR PACKINGS

It is reported to be preferable to use experimentally determined F_P values for the packing rather than F_P calculated from packing geometries. This is really indicating that the correlation is not very accurate, and the accuracy is improved by fitting the correlation to experimental data for each new packing. Prediction of the flood point is not very accurate, and some authors warn that a $\pm 30\%$ error is to be expected.

Having chosen the packing type and obtained its packing factor F_P the expression corresponding to x axis of Figure 7.9:

$$\frac{L'}{G}\sqrt{\frac{\rho_g}{\rho_L}}$$

can be calculated. Figure 7.9 is then used to read the term on the y-axis corresponding to flooding;

$$\frac{v' \, F_p}{g}\left(\frac{\rho_g}{\rho_L}\right)\left(\frac{\rho_{H_2O}}{\rho_L}\right)\mu^{0.2}$$

[Note that each of these terms is dimensionless (g is the gravitational constant, 9.81 m s^{-2} in SI units), with the exception of viscosity (μ) which is in centipoise. Note also that the liquid and vapour flow rates, and densities are required in weight, not molar units. Finally, ρ_{H_2O}/ρ_L is a correction introduced by Eckert for liquids other than water. ρ_{H_2O} is the density of water in units consistent with the other density units.]

Since all the components of this expression are known with the exception of v', the maximum vapour velocity at the point of flooding (m/s), then this can be calculated.

Hence, the column diameter for, say, 80% flooding can be calculated:

$$D = \sqrt{\left(\frac{3600 \ G}{0.8 \ v' \rho_g}\right)\frac{4}{\pi}} \qquad\qquad (7.22)$$

where G is the actual weight gas loading required to be handled by the column.

Having calculated the column diameter, a check should be made that the pressure drop is acceptable and that the liquid wetting rate is not too low.

Figure 7.9 can be used to determine the resulting pressure drop at gas rates below the flood point (e.g. 80% of flooding). With this pressure drop/column height, once the height is determined, it is possible to confirm that pressure drop is no problem - or modify the design if it is. Such modifications could be changes in the packing type, or working at lower vapour loadings.

HEIGHT

The distillation calculation has determined the reflux ratio and number of theoretical plates (N). The packed height is determined from the number of theoretical plates via the HETP:

$$H = n \ x \ HETP \qquad\qquad (7.23)$$

The HETP, as discussed at the beginning of this chapter, is a function of the system and the packing, but primarily the packing. The smaller the packing the greater is $K_g a$, and the smaller is the HETP.

As a general rule, the HETP is related to the characteristic nominal diameter of the packing (d (m)) by the expression:

$$\text{HETP} \simeq \frac{d}{0.075} \quad (m) \qquad\qquad (7.24)$$

Table 7.5 contains some HETP values for various types of packing when distilling fairly difficult-to-separate organic systems.

The HETP of course is dependent on many factors: flow rates, entrainment and maldistribution. Figure 7.5 shows a typical relationship between HETP and column loading. At low rates the HETP will rise because of wetting problems, and at high rates entrainment will again cause the HETP to rise. These two extremes define the operating range of the column.

REDISTRIBUTION

To obtain the quoted HETP for a packing, the liquid must be well distributed over the packing initially, and then it has to be redistributed after a certain length since the liquid migrates to the walls and becomes mal-distributed.

This has to be done with redistribution devices placed at frequent intervals in the column. Distributors between 2 and 4 meters are reasonable. These devices can take the form of a total liquid collection and redistribution (a complex device as expensive as a bubble tray); or it can be a simple ring around the wall to partially redistribute the liquid collected at the wall; or, alternatively, it may be one and then the other, to achieve a reasonable redistribution at a reasonable cost.

The redistributors themselves must be designed, or supplied designs of distributor must be checked. Figure 7.10 shows the basic structure of these distributors.

The vapour collection must not involve too high vapour velocities, and the liquid is preferably distributed from slotted tubes to reduce problems of blocking and incorrect levelling associated with a simple perforated plate. The number of distribution points and the tube weir dimensions must be chosen so as to handle all expected liquid flows. This involves using the weir formula given as Equation (7.20) to check that there is an appreciable head at lowest rates, and that at highest rate the weir is not flooded.

Packing supports must also be specified. These should be a form of a grid, small enough to prevent the packing falling through, and with sufficient free area so that it does not form the bottleneck and flooding point of the column.

Commercially designed grids are supplied by packing suppliers. Sometimes larger packing is used directly on the grid to prevent flooding at this point. The specified packing is then put on top of the initial layer.

Packing hold-down plates are sometimes specified above each packing section. These again take the form of grids and are there to prevent the packing being shot out of the column sections during periods of instability and flooding. It is not unknown to find large quantities of packing in the condenser of a packed distillation column!

Having chosen a packing type and dimensioned the resulting column, consideration should be given to alternative packings with improved characteristics if any problem show themselves in the design (e.g. low wetting, pressure drop, excessive height, etc.). Other possibilities should be followed through to see whether cost reductions are possible. Many different packings are available, and a cost analysis is often necessary to make a final considered judgement.

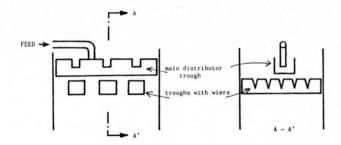

(a) TROUGH DISTRIBUTOR (LARGE DIAMETER TOWERS)

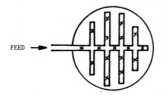

(b) PERFORATED PIPE DISTRIBUTOR

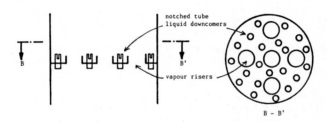

(c) REDISTRIBUTOR

FIGURE 7.10
LIQUID DISTRIBUTORS AND REDISTRIBUTORS FOR A PACKED COLUMN

7.5 WETTED WALL COLUMNS

Wetted wall columns are normal diameter columns with no internals, they have very low pressure drop, but also very high HETPs, and thus separations requiring many plates cannot be considered with them. They do, however, find use in high vacuum separations, because of their good pressure drop /theoretical tray characteristics.

Table 7.5 gives a comparison of pressure drop/plate and HETP for the various column packings used when low pressure drop is required.

A second feature of wetted wall columns is their low hold-up, and this makes them suitable for temperature sensitive products (specific organics, foodstuffs, etc.). There remains the problem of the thermal degradation in the boiler, and this is solved by having a film vaporiser in place of the pool of boiling liquid as already discussed in Chapter 6.

The commercially developed equipment for such situations is wiped falling film evaporator (Luwa), which has rotating internals that wipe the heated surface, thus improving the heat transfer coefficient and reducing the film thickness. When the rotor that wipes the surface is cooled, fractionation occurs in the column in addition to simple vaporisation. It is usual to build two sections: a thin film vaporiser feeding a thin film rectifier (both heating and cooling). Thus we have a distillation system with very low hold-up. The system is run at sub-atmospheric pressure to improve product stability, and this gives the general form of equipment in which wetted wall column distillation usually finds application.

TABLE 7.5

COMPARISON OF DIFFERENT COLUMN INTERNALS FOR PRESSURE DROP

INTERNALS	LOADING F FACTOR $(kg^{1/2}m^{-1/2}s^{-1})$	SPECIFIC PRESSURE DROP (mbar/theoretical plate)	HETP
25 mm Pall rings	1.0	0.7	
	2.0	1.8	
80% loading	2.3	2.5	.33
	2.5	3.0	
50 mm Pall rings	1.0	0.5	
	2.0	1.5	
80% loading	2.5	2.4	.46
Sulzer-type packing	1.0	0.06	
	2.0	0.3	
80% loading	2.8	1.4	.27
	3.0	1.8	
Thin film columns	1.0	0.15	
	2.0	0.4	
	3.0	0.6	1.0
	4.0	0.8	

7.5a DESIGN PRINCIPLES

Systems suitable for wiped thin film columns are generally so individual that no VLE or performance data exist. VLE, viscosities and quantitative analysis of temperature sensitivity must generally be determined by experiment, thus test runs with manufacturers' pilot plants are the normal procedure. The continual vaporisation from the walls and condensing on the wiper blades causes a flow that is difficult to equate to a number of theoretical stages and reflux ratio, and thus heat flow requirements cannot be directly developed from a McCabe-Thiele or normal distillation calculation.

The thin film fractionator is not suitable for columns requiring many plates. The HETP is approximately the same as Pall rings, but the column height is limited by the bearings and supports of the rotary internals that wipe the walls.

For orders of magnitude calculation heat transfer coefficients of 2 kW $m^{-2}K^{-1}$ can be assumed, and HETP's of about 1 m. Superficial loading velocities are high, allowing velocities of 3.5 m/s to be used with system pressures of the order of 20 mbar.

Maximum column sizes are about 10 m long and 1 m diameter. They are, of course, because of their mechanical internals, expensive pieces of equipment.

For accurate design work, the manufacturers must be contacted. They have correlations for their equipment, past experience, and will be able to say whether an application can be designed on paper or whether trials will have to be run.

7.6 CAPITAL COSTS

The estimation of the total capital costs of distillation equipment is, as with the cost estimation of other equipment, based upon obtaining the main item equipment base or "fob" cost, which is multiplied by a factor to obtain a cost which covers the purchase plus installation. This factor makes allowance for all costs; foundations, piping, instrumentation, painting, insurance, design etc.; it is based on the analysis of achieved past costs, and is generally between 3.0 and 4.0.

Distillation equipment main items are columns, heat exchangers, vessels and pumps. Numerous correlations of cost information exist in the literature because cost information outdates quickly and so it is always necessary to use recent data to obtain reliable costs. However, for the sake of illustration we are not constrained to recent publications, and so we can demonstrate costing with an older, but complete method. In 1969 Guthrie published a very complete set of data which is good for demonstration purposes. An estimate of present-day costs can be obtained by using the inflation factor (f_I), though of course, the bigger the inflation factor, the greater the inaccuracy. To bring 1968 $US data to equivalent 1984 $US values, an inflation factor of the order of 2.5 should be used.

Column

Column "fob" costs are composed of the shell (or vessel) cost, which is defined by its diameter and length, and the cost of the internals. Shell costs are dependent on materials and construction and working pressure and be defined by the two factors f_M and f_P by the following equation:

$$\text{Shell item costs} = f_P f_M (C_B)_{shell} \qquad (7.25)$$

where: C_B = base "fob" costs (1000 $ 1968)
        f_M = material factor
        f_P = pressure factor

Values for these two factors are listed in Table 7.6 . C_B is given by the empirical equation

$$(C_B)_{shell} = \exp(1.33F' - 0.541) \qquad 1000 \ \$ \ (1968) \quad (7.26)$$
where;

$$F' = (0.778 - 0.000082L) \ \ln(3.281D) + 0.9199 \sqrt{L} - 1.433 \quad (7.27)$$

D(m) is the vessel diameter and L its length (or height) (m).

Internal costs are then added to produce the complete column "fob" cost. If the internals are plates, then the cost of a plate is correlated to diameter. Factors are then applied for tray type, materials construction and tray spacing. The cost of the trays in a column can therefore be correlated by the following expression:

$$(C_B)_{tray} = 0.030 + 0.038D^2 \qquad 1000 \ \$ \ (1968) \quad (7.28)$$

$$\text{Total tray item costs} = N(C_B)_{tray} (1 + f_T + f_M) \qquad (7.29)$$

Column packing are usually quoted in price/m ^{3} volume. Each packing type has a specific price. Various packing prices, scaled to the $(1968) prices are given in Table 7.7.

Hence for a packed height h and diameter d, the packing costs are given by:

$$\text{cost of packing} = \frac{\pi}{4} D^2 L (C_B)_{\text{packing}} \tag{7.30}$$

The total column "fob" cost is then the addition of the shell to the tray or packing costs.

TABLE 7.6
FACTORS REQUIRED BY THE COST ESTIMATION PROCEDURES

TABLE 7.6a - COLUMN SHELLS and VESSELS

Pressure (bar)	f_P	Material	Clad f_M	Solid
up to 4	1.00	carbon steel	1.00	1.00
8	1.05	stainless steel	2.25	3.67
15	1.15	monel	3.89	6.34
20	1.20	titanium	4.23	7.89
40	1.60			
70	2.5			

TABLE 7.6b - TRAYS

Tray Type	f_T	Material	f_M
sieve	0.0	carbon steel	0.0
bubble cap	1.8	stainless steel	1.7
valve	0.4	monel	8.9

TABLE 7.6c - HEAT EXCHANGERS

Exchanger Type	f_T	Pressure (bar)	f_P	Material	f_M
fixed tube plate	-0.2	up to 10	0.0	carbon steel	0.0
floating head	0.0	20	0.10	stainless steel	2.0
kettle	0.35	40	0.4	monel	3.0
		60	0.55	titanium	10.0

TABLE 7.6d - PUMPS with MOTOR

Material	f_M
cast iron	0.0
stainless steel	1.9
monel	3.2
titanium	9.0

TABLE 7.7
CAPITAL COST OF COLUMN PACKINGS

- Converted to $(1968) to be consistent with Guthrie's data.

Nominal size (mm)	25	35	50
Pall Rings (stoneware)	180	120	110
Pall Rings (carbon steel)	350	240	218
Pall Rings (stainless steel)	1341	1024	882
Pall Rings (polypropylene)	318	215	198
Saddles (stoneware)	264	194	180
Sulzer Structured (stainless steel)	approx	3000	

Exchangers

Exchanges (boilers, condensers, preheaters) are all correlated by the same expression, based on heat exchange area. Factors are used to correct for exchanger type, materials of construction and pressure. The following equations yield the exchanger base and fob costs.

$$(C_B)_{\text{heat exchanger}} = 0.73 + 0.30A^{0.65} \qquad 1000 \; \$ \; (1968) \qquad (7.31)$$

$$\text{heat exchanger item cost} = (C_B)_{\text{heat exchanger}} (1 + f_T + f_p + f_M) \qquad (7.32)$$

Horizontal Storage Tanks

The cost correlation for horizontal pressure vessels can be used to determine the cost of feed, product and pumping tanks, in the system. The correlation is the same as for vertical pressure vessels used for the column itself, using the same equation (7.27) to calculate F'. However, in place of Equation (7.26) a modified form (Equation (7.33)) is used reflecting the lower price of horizontal vessels. Equation (7.25) allows for pressure and material changes from the base cost.

$$(C_B)_{\text{vessel}} = \exp (1.29F' - 0.969) \qquad 1000 \; \$ \; (1968) \qquad (7.33)$$

Pumps

Pumps are comparatively minor cost items, which can be correlated to the following equation:

$$(C_B)_{\text{pump}} = 0.35 + 0.073 \sqrt{Q\Delta P} \qquad 1000 \; \$ \; (1968) \qquad (7.34)$$

this base cost is then corrected for material as follows:

$$\text{pump item cost} = (C_B)_{\text{pump}} \, f_M \qquad (7.35)$$

The value of f is given by Table 7.6. The resulting costs applies to low pressure centrifugal pumps including electric motor drive.

The total installed cost of a distillation section of a plant is therefore the sum of the "fob" costs multiplied by the factor to convert from "fob" item cost to total installed equipment cost, including all associated costs including design, transport, installation etc. For this type of equipment the factor is 4.0. The final cost has to be corrected for inflation since 1968 using inflation factor f_I.

$$\text{Total installed cost} = \Sigma(\text{Item costs}) \times f_I \times 4.0 \qquad (7.36)$$

There are many variations with different degrees of detail - the "fob" to install costs being individual for each item of equipment, and proportional to equipment cost; inflation correlation separated into labour and materials, etc., but for purposes of instruction the above method suffices.

The full Guthrie method is available in program form as the EURECHA program CAPCOS. Further details and information concerning its availability are given in the Appendix on Eurecha programs.

7.6a MARGINAL COSTS AND COST OPTIMISATION

Engineers are interested in costs in order to "optimise" their designs, to find a suitable balance between capital and running costs. It is necessary to suggest a number of alternatives, and compare the cost of these alternatives to make a decision. Unfortunately, published capital cost methods, including the one above, are based on allocating the overhead cost proportional to the main equipment "fob" costs; hence the use of factor of 4.0 to obtain total cost from equipment costs.

Now clearly the addition or removal of some trays in a column is not going to affect the building costs, the design costs, and probably not even the equipment foundations. For analyses of alternatives involving minor changes, the standard capital cost method will over-estimate the effect of the equipment cost changes. We must work with "marginal costs". These marginal costs will be somewhat more than the bare difference in "fob" costs, because extra costs will be involved when equipment sizes change, and so some factor between 1.0 and 4.0 would best reflect marginal cost changes. Very little work seems to have been done on the real marginal costs of equipment changes, and in the absence of better guidance one could take 2.0 times the "fob" cost changes for optimization studies.

In practice the procedure is even more complex because so many design decisions are "integer" in that one design with one cost cover a range of capacity, beyond which a different design with a step change in cost is involved, which again covers a further range without a cost increase. This means that towards the end of a design exercise the design should be modified to fully utilize the ranges presented by the standard sizes that have evolved. We are right back to Chapter 1, and the uneven Pareto surface of Figure 1.3, trying to move to the local apex representing an improvement in a performance index with minor changes in the costs.

The inexperienced engineer must watch against the old cost engineer's habit of taking every opportunity of increasing his estimate, and never admitting that a proposed change could actually reduce costs:

"Saving a few plates doesn't reduce the capital costs - it's only a few sheets of metal perforated by machine - the saving is insignificant", is the cost engineer's reply to an enthusiastic engineer who finds he can develop a better plate efficiency.

"The costs would rise significantly, new foundations - the increased
windage you know; extra platforms, we would have to change the structure
supporting the condenser, it is unfortunately going to be very expensive
to add these two trays", is the reply when close study reveals more plates
would bring savings in running costs due to the low reflux ratio.

Of course the cost engineer is often right, and design engineers usually have
an inflated impression of the cost savings their new ideas would bring. If the
designer engineer had his way he would quite quickly whittle away the capital
cost down to zero by a series of cost reduction schemes, unless the cost
engineer was there to bring a little realism into the situation.

7.7 FURTHER READING

Billet, R., 1973. Industrielle Destillation, Verlag Chemie, Weinheim.
                   - full of practical data and detail.

Guthrie, K.M., 1969. Chem. Eng., March 14th, p114.
                   - complete capital cost estimation method.

Holland, C.D., 1981. Fundamentals of Multicomponent Distillation, McGraw-Hill.
                   - details of plate efficiency predictions.

Perry,R.H. and Chiltern, C.H., 1973. Chemical Engineers' Handbook, McGraw-Hill.
                   - good survey of gas-liquid contactors, full of detail.

7.8 QUESTIONS TO CHAPTER 7

(1) What is the role of the column internals?

(2) Why are plate efficiencies for different distilling systems and different
    plate designs quite similar?

(3) What factors govern the choice between plates and packing?

(4) What dimensions must be given to fully specify a distillation tray, and how
    are each of these determined?

(5) What factors govern the choice of packing type?

(6) When are wetted wall columns useful?

Chapter 8

SAFETY AND CONTROL

The equipment is designed. Now it must operate under the conditions the designer intended and also operate safely. This is achieved by specifying a suitable control system and alarms, and by defining the correct pressure relief equipment.

8.1 CONTROL

The basic objective of the control scheme is to ensure that the column operates as defined by the design. A more advanced objective is that, in addition to this basic requirement, compensation should be made for deviations from the design condition; energy can be saved by careful trimming, and sufficiently robust control schemes can be derived to achieve stable control over wide ranges, allowing start-up and production level changing to be satisfactorily accomplished automatically.

8.1a THE BASIC CONTROL SCHEME

The column design specifies that, for a certain feed rate and composition, a certain overhead flow must be taken from the column and a certain reflux returned to the column; the boiler must produce enough vapour to satisfy these flows, and then the required separation will be achieved.

There is a mass balance and a heat balance to fulfill, and a reflux requirement to maintain to achieve the separation.

The simplest control to achieve this is to run with a controlled feed rate, take off top product at a controlled rate, and supply sufficient steam to the boiler to attain the reflux rate specified by the design.

This control system has a very serious disadvantage. There is no feedback on the distillate rate removal. This is dependent on feed rate and feed composition, and changes of these variables will require manual changes to the distillate flow rate setting. Also, the demand on the accuracy of this feedback for the mass balance for most distillations is extreme. If a column separates to 0.01% levels, top and bottom, the mass balance must also be accurate to 1 part in 100,000. This is impossible to achieve without some form of feedback. The only exception to this rule is the separation of mixtures into only a few fractions, such as in oil refining, where satisfactory control can be achieved by setting the off-take rates for the various top and side products. In this case, purity is not the criterion, but only splitting the feed into a series of fractions. Since very many components are involved, the shift from one component to the next does not produce a marked change in product quality. This is in contradiction to the binary distillation case, where any shift from one component to the other "to achieve a mass balance" is disastrous to the product quality.

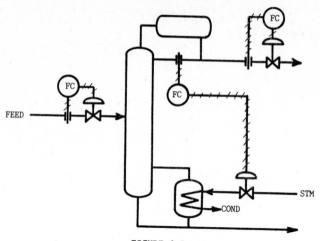

FIGURE 8.1
COLUMN CONTROL BY FLOW CONTROL ALONE

Mass Balance Feed-Back Control

The feedback on the distillation take-off rate must be achieved by measuring
the composition in some way. Given a measurement of product quality at one end
of the column, and given the column operating with the designed reflux,
the top and bottom products will be in specification, and the feedback control will adjust to the distillate rate to achieve the mass balance to a high
degree of accuracy. (This will be upset by feed composition changes. But remember that a good flexible design contains enough safety factor to cope with expected fluctuations in feed without modifying operating conditions.) Figure 8.2
shows this arrangement.

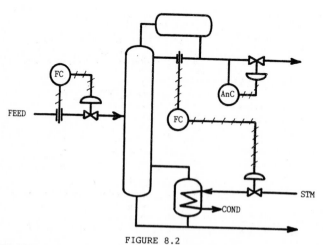

FIGURE 8.2
SIMPLE FEED-BACK CONTROL ON PRODUCT QUALITY TO ACHIEVE MASS BALANCE CONTROL

The possibilities open for measuring product quality are very limited and expensive, involve delicate instrumentation requiring much high quality maintenance, and have a slow response. It is therefore more usual to use a secondary
response that is a function of product quality. This response is temperature.
Since we are dealing with products with different boiling points, the temperature is a good measure of composition. This results in the temperature profile
within the column (see Figure 8.3).

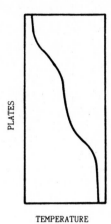

FIGURE 8.3
TEMPERATURE PROFILE FOR CONTROL PURPOSES

Measurement of product quality from its boiling temperature is generally not
possible because at high purities the difference in temperature between on- and
off-specification material cannot be observed, because of the accuracy of the
temperature measurement (and total pressure control). However, as we move down
the column, sensitive temperature regions can be found, and if we maintain our
reflux or L/V ratio in the column at the design rates, this fixes the top and
bottom qualities.

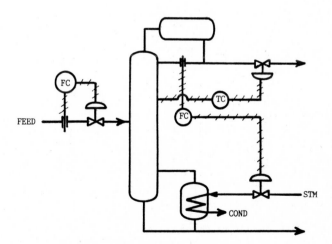

FIGURE 8.4a - Temperature profile controlled by distillate take-off

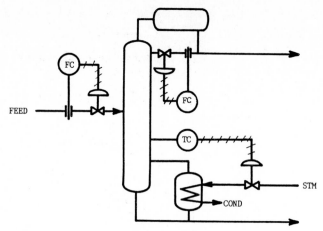

FIGURE 8.4b – Temperature profile controlled by steam

FIGURE 8.4
BASIC COLUMN CONTROL SCHEMES

Hence, the most used detection method for product quality is by temperature measurement at a carefully chosen point within the column. The control scheme then develops to that shown by Figure 8.4a. Figure 8.4b is an equivalent control in that the reflux rate is flow controlled, and the temperature profile in the column controls the steam fed. Scheme 8.4b will have advantage of speed of response over 8.4a, since the steam will have an earlier response part way up the column than at the top of the column. The distillate rate is then simply the remaining condensate not required for the reflux. The column profile is held at the correct point, the reflux is correct, and therefore the mass balance will be achieved and the products will be in specification.

The control shown in 8.4b operates with a fixed reflux flow. At lower feed rates than design, the fixed reflux flow represents an excessive flow and improved product quality but is expensive in energy and "quality give-away". The separation requires a fixed reflux ratio, and therefore a design that relates the reflux rate to the distillate rate results in energy saving at lower feed rates. This is shown in Figure 8.5. (It is also possible to fix the reflux/feed ratio, which might be more efficient if the bottom product quality is the more important). This modification also reduces the need for feed flow control, because the reflux ratio control will make the necessary compensation for feed flow fluctuations.

Pressure Control

The pressure in an atmospheric distillation is usually governed by the vent line from the condenser that is left open to the atmosphere. Hence, the top of the condenser is effectively controlled at atmospheric pressure. For distillation at other pressures, the same technique can be used by fitting a pressure control valve in the condenser vent line and controlling the operation of this valve by the condenser pressure.

This control procedure functions by controlling the amount of inerts in the system. The inert "blankets" part of the condenser area, thus effectively controlling the amount of condensation that occurs and it is this that governs the pressure attained in the system. Hence, if a feed does not contain inerts, or contains very little inerts, the system cannot function, or will function

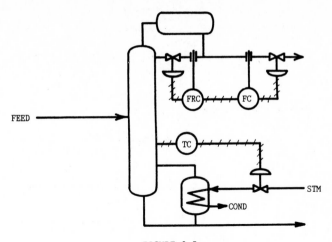

FIGURE 8.5
REFLUX RATIO CONTROL

only very slowly if the inerts are to be supplied by the feed. The problem is
easily overcome by feeding a bleed of inerts immediately before the pressure
control valve. The control valve is then operating on this inert stream, and
the necessary speed of response of the controller can be changed by adjusting
the flow of this inert bleed. This control system is standard for vacuum operation, and can be used for pressure operation when inerts are not present. In
refining of petrochemicals, sufficient H2 and other lights are usually present
in distillation systems and thus the bleed becomes unnecessary.

Level Control

The control so far developed ensures that the mass and energy balances are
maintained and pressure is controlled, but the bottom product collects in the
boiler. This has to be removed from the column, and this is usually done by
supplying a simple level control system in which the liquid level in the
sump is controlled by the bottom liquid take-off from the system.

There also is the practical problem associated with collecting the liquid from
the condenser and distributing it as reflux and top product. The condenser
must not be used as the liquid reservoir, because this blankets the tubes and
causes extra problem (e.g. pressure control difficulties). The liquid must
therefore be collected in a reflux tank, and this then requires a level control
system with its level controlled by the flows leaving the tank.

A level controller can control the flow of the largest stream and the smallest
stream related by ratio control to the larger stream, as shown in Figure 8.6.

Figure 8.6 is a good standard control system for a distillation column requiring pure components at each end. Response times are low in the pressure and
temperature profile control loops. Heat is economized and feed flow variations
accounted for by the reflux ratio control of reflux.

This is not the only way of controlling a distillation column, and Rademaker et
al (1975) mention a further 47 schemes! Ideally, there are 5 control loops associated with distillation control, giving a total of 5 x 4 x 3 x 2 = 120 different possible systems. Forty-eight of these will function, and many of the 48
are the best for particular cases.

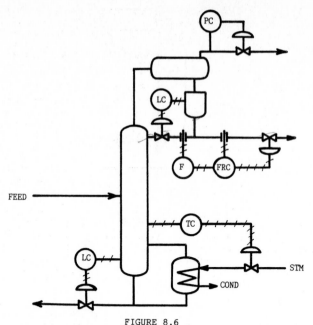

FIGURE 8.6
STANDARD CONTROL SYSTEM FOR DISTILLATION SYSTEM WITH PURE COMPONENTS

8.1b OTHER CONTROL SYSTEMS

There is no systematic way of defining which control system is best for which
distillation system, and the best approach is probably evolutionary. Figure 8.6
shows a good method for a standard distillation. In which way does an individual distillation differ from the straightforward case, and what changes does
this suggest for the control scheme? Some cases of deviations from the standard
distillation and the appropriate control schemes are given below. It is not a
complete summary, and each distillation must be considered as a new problem.

(a) <u>Non-pure products with very many components</u> e.g. oil refining. In this case
temperature profiles are less sensitive and mass balances much less critical.
Hence a control scheme devised for taking flows out of the column as various
products at control rates (or possibly flow rate controlled to the feed) and
provision for maintaining reflux rate are more suitable than temperature profile control. Of utmost importance in oil refining is the minimisation of energy costs.

(b) <u>Close boiling point with difficult separations</u> usually result in column
pressure drop masking the composition influence on the temperature profile. Simple temperature sensing is then no longer effective. Measures to be taken are
to relate the temperature measurement to the pressure, or choose a control system that is not dependent on temperature measurement (e.g. analysis).

If the pressure sensing point for the pressure control is moved to the temperature sensing point, the temperature recorded reflects the composition profile (at the expense of a less direct pressure control).

If pressure is measured with a temperature, a small microprocessor can make
the pressure correction to temperature to again produce a signal that is a
function of composition.

A vapour pressure bulb rather than a temperature sensor can be used in the column. This consists of a bulb containing the liquid at the composition to be attained at that point in the column. Then the pressure difference between the column pressure at that point and the vapour pressure in the bulb is a measure of the deviation of the composition from that required. The idea is good, but has never been enthusiastically reported on, because of practical problems in its installation.

(c) Either top or bottom product is disproportionally small. It is not advisable to use small flows for control purposes, because the effect they have on the system is slow or even inadequate. If a control scheme is developed where low flows are involved, care must be taken to see that a major control requirement, such as the mass balance over the column, is not achieved by manipulating a small flow. The standard control scheme is not affected by low distillate flows because the distillate is ratioed to the reflux flow and is, therefore, in effect, not a low flow, and the boiler flow is only to maintain a bottom liquid level, which is not critical.

(d) Quality of only one product is of importance. When only one product is of importance, there is an opportunity to develop a control scheme more suitable for achieving a constant quality of the important stream at the expense of the second stream. This may involve considering a new control scheme (to maintain one quality and maximum loading in place of two qualities for example) or it may be concerned with the adjustment of the controller settings to obtain a tight control on the important quality at the expense of a slack control of the unimportant one.

Theoretically speaking, the relaxation of one of the constraints allows for an on-line optimisation to be of advantage to minimise overall operating costs. This involves finding the economic optimal composition for the unspecified stream. Such systems have been proposed but not implemented on an industrial scale.

(e) Top product is recovered as vapour. Sometimes it is necessary or required that the top product is taken from the column as vapour. This is necessary if the top product is very volatile and has to be separated from condensable impurities. In such a situation, the condenser temperature level can sometimes be so chosen that the product remains non-condensable, and the reflux condenses out, no control whatsoever being required, as long as excess coolant flow is allowable.

In other situations, the top product is required as a vapour and, rather than involve the extra heat loads and equipment to condense and revaporise it, it is best extracted from the column as a vapour. In this case, it is the duty of the condenser to condense only the reflux. Such a system is highly dependent upon control for its operation, and there is the interaction between column pressure and take-off rate (and hence mass balance), which makes for further difficulties. It is now necessary to control the coolant flow, which is a slow response system, and control the pressure, which can be fast.

A reasonable solution is to control pressure by vapour product removal, control reflux to the column to maintain a fixed reflux ratio (or fixed reflux) and use the condenser coolant flow to control the condensate pumping tank level.

If the feed consists of very many components, as in oil refining, the mass balance criterion is not so critical, and this makes it possible to flow-control the vapour take-off, using coolant to control column pressure.

If a dephlegmator is used, mounted on the top of the column, and the reflux and product have similar condensing temperatures, the reflux must be determined by heat balance over the coolant circuit, and the coolant flow regulated to obtain the required reflux or reflux ratio. This would require a microprocessor to make the necessary calculations and conversions.

(f) <u>A control system to minimize the use of condenser coolant</u> All discussion so
far has assumed that maximum coolant flow is set to the condenser and the
condenser heat load control is determined by the inerts blanketing of the
condenser tubes. In a recycling condenser coolant situation this usually has
little economic disadvantage, but in a non-recycling system with high coolant
costs there is a potential for cost savings by restricting the coolant flow and
allowing the maximum temperature rise to occur in the coolant.

From a control point of view there are problems. The highly non-linear relationship between flow and heat flow demands a high rangeability for the control
valve, and the fixed maximum temperature of the coolant leaving are difficult
problems to solve. Control of condenser coolant flow is worth considering only
if it is confirmed that there is a real saving in reducing coolant flow, and
that the saving is appreciable. If this is the case, the system pressure would
in the first place be controlled by coolant, with possibly a switch to inert
control when the above mentioned problems present difficulties.

(g) <u>A system to maximise column capacity</u>. The column maximum capacity is most
directly affected by the vapour loading. If a column has become a bottleneck to
production, and it is important that it operates on the point of flooding, then
it is no longer optimal to use vapour to control the temperature pattern, because this requires running below maximum vapour load to leave flexibility for
control. In such a case, the vapour load could be controlled at maximum (e.g.
by pressure drop across the packing controlling the steam to the boiler) and a
control scheme then developed with this as the basis.

8.1c PROBLEMS TO AVOID IN CONTROL SYSTEMS

So many control possibilities exist that there is no easy way of discussing
each one. However, some of the problems associated with poor control loops can
be indicated. These problems do not disqualify a particular control system, but
they make it more difficult, and its other advantages must therefore be
correspondingly more attractive.

(a) <u>Slow response and dead-time</u> The pressure should have a response time of
seconds, the temperature profile in minutes, and the level control in tens of
minutes. When these times are extended, problems occur. Controlling pressure
by the level (flooding the condenser or boiler) is bad. Controlling bottom
temperatures by top reflux flow is worse than controlling top temperature by
top reflux flow. Controlling level by steam to the boiler is poor. But, though
these controls are poor, they may be necessary in particular circumstances.

(b) <u>Inverse response</u> Responses that can change sign within the control range
are not advised. Temperature differences between two points in a profile overcome total pressure and pressure drop problems associated with relating temperatures to composition profiles. Unfortunately, when the profile is either too
high or too low, the temperature difference is zero. One can work on one
slope of the response curve, but if perchance one finds oneself on the other
side the control action is in the wrong direction (see Figure 8.7).

A similar inversion can occur using coolant flow to control pressure. A reduced
coolant flow increases the pressure but also reduces the reflux subcooling. If
the internal reflux is so reduced that heavy components appear at the top of
the column, the temperature driving force of the condenser is increased,
and this increases the heat transferred which results in a pressure reduction -
the inverse of the required effect.

(c) <u>Coupled control loops</u>. If control loops that act on related variables are
proposed, coupling of loops can occur and this results in instabilities and
difficult control situations. A coolant flow/column temperature, inerts
flow/column pressure could be coupled, producing an oscillating system.

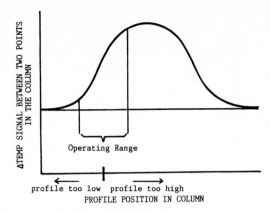

FIGURE 8.7
INVERSE RESPONSE OF TEMPERATURE DIFFERENCE CONTROL

Systems are sometimes proposed that control top and bottom purities by two temperatures, one near the top and the other near the bottom of the column (see Figure 8.8). Such a system is an attempt at optimal control, where the operating reflux rate is determined on-line rather than the designer's specified ratio (or a modification of this ratio in a light of operating experience) being accepted.

Such a system has the potential of minimum operating costs and maximum production capacity by reducing quality give-away, but the cost is the introduction of a highly coupled control system. Before such a proposal is accepted, it must be carefully checked for the effect of total pressure fluctuations and pressure drop at various loadings. It is likely to be useful only for wide-boiling, easily separated systems.

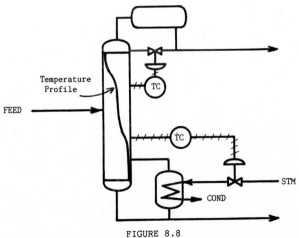

FIGURE 8.8
TEMPERATURE CONTROL OF BOTH STEAM AND REFLUX

8.1d IMPROVING STABILITY

Despite reports in the research literature that oscillating mass transfer systems operate more effectively than steady systems, it is the aim of every operator to have the column operating absolutely steadily - no fluctuations in product quality or column loading.

The control scheme serves two purposes. Firstly to adjust conditions to those that will produce on-specification product - which is what we have been discussing - and secondly, to compensate for the external fluctuations in order to maintain steady operation in the event of these changing external influences.

The external influences are from:

- changing steam main pressure owing to other uses in the network. Changes can occur in seconds or minutes.

- changes in feed composition because of operation of up-stream plant. Feeds are usually fed from tanks with 2 to 12 hours capacity, resulting in feed fluctuation being in hours rather than minutes.

- changes in feed flow rate. This may be level controlled from up-stream plant - again variations with cycle times of hours.

- changes in condenser cooling temperature, with a daily cycle.

- changes in column heat losses, which occur in minutes when rain storms occur.

The control systems so far proposed (Figure 8.6 for instance) cope with the slower changes, because once the fluctuation has affected the column operation the control system will bring it back to specification. The faster fluctuations will not be compensated for but will simply introduce disturbances, which follow through the whole system.

The situation can often be improved by using cascade control where the short-term instabilities are removed with a quick response flow control loop, and the longer term control action is used to adjust the set point of the flow controller. Figure 8.9 shows the control system of Figure 8.6 using full cascade control.

Such complete cascade control increases the control costs considerably, and in many cases there is no need for the flows to be steady; or very steady control can be achieved by setting a very slow integral action and having sufficient capacity.

Hence, bottom outlet cascading control is not necessary if the flow is to a bottoms product tank; cascade of reflux is necessary only if the capacity of the reflux tank is low (i.e. less than 15 minutes), and cascade of the distillate is not necessary if it discharges to a product stage.

Operating experience is necessary in order to judge where real advantages are gained by paying for cascade control.

8.1e FEED-FORWARD SYSTEMS

Feed-forward control systems are the ultimate aim of the control engineer because it means that corrective action can be taken before the process is disturbed. This gives improved quality control, closer operation to maximum rates, and less quality give-away. Feedback control requires process disturbances to occur for the corrective action to be determined.

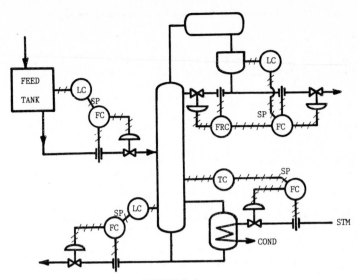

FIGURE 8.9
BASIC DISTILLATION CONTROL WITH FULL CASCADE CONTROL FOR IMPROVED STABILITY

Disturbances in distillation come from feed flows, compositions and enthalpy.
In situations where feed flow cannot be controlled - within a process sequence
without intermediate storage, for instance - any flow control loops can be
ratioed to the feed flow. Such candidates as steam flow, or product take-off,
could be so ratioed, cascaded as necessary.

Feed-forward for the other disturbances (feed enthalpy and composition) are
rather more difficult to achieve. The feed enthalpy measurement and corrective
actions are not simple, and feed composition measurement for a multicomponent
system is also a difficult task. Research papers describe systems, but they are
not yet applied on a commercial scale.

8.2 ALARM SYSTEMS

Having determined the control system, the total instrumentation has to be
specified : which variables are indicated and which recorded, and where should
alarms be included.

There needs to be provision for seeing the past history of the control
variables and the controlled variables. The temperature at the control point,
the reflux rate or ratio, the pressure (unless it is atmospheric) and the feed
rate all need to be recorded (or the recorded retrievable from the computer
storage) to reconstruct periods associated with off-specification operation or
unstable operation.

Alarms should be specified to inform of malfunctioning that will result in off-
specification material or in equipment damage. A concept needs to be worked out
defining which variables have advisory alarms, which belong to a more serious
category, and which are so serious that they demand immediate automatic action.
System pressure (high), cooling water flow (low), pumping tank levels (high and
low) would be candidates for alarms.

8.3 SAFETY

8.3a FIRE RISK

Since distillation is very often concerned with handling volatile organic compounds, inflammability is probably the most serious hazard associated with distillation equipment.

When any equipment is concerned with the handling of inflammable liquid, the area must be designated as flame-proof, excluding all possible sources of ignition, so that, in the event of a spillage, the material can dissipate before it ignites.

A flameproof area entails all electrical equipment being explosion-proof. This involves all electrical equipment, from switches to motors, being encased in very solid, well-fitting housing, so that any explosion occurring within the electrical equipment is not transmitted outside it.

An alternative technique for electrical equipment is to involve such a low voltage that sparking is not possible. 'Intrinsically safe' control equipment can be so specified. Low power electrical signals are perfectly adequate, but such a technique cannot be used for any electrical equipment requiring power – for motors or lighting, for instance.

Flameproof areas also involve access restrictions on all forms of transport capable of being a source of ignition.

8.3b OVERPRESSURE

The most likely operating hazard associated with distillation is the overpressure that develops when the condenser system is restricted.

The boiler is capable of generating large quantities of vapour, which are handled by condensation at the top of the column. Should this condenser fail (e.g. because of coolant failure), the vapour will increase the pressure in the equipment and there have to be provisions for handling it.

The two possibilities for handling a condenser failure are:

- to design a whole equipment to withstand the vapour pressure of the components at the maximum heating medium temperature

- to provide equipment for relieving the pressure once it reaches the design pressure of the equipment. This involves a relief system and provision for handling the released vapours.

Since it is usually impractical to design for the maximum allowable system pressure, it is usual to supply relief equipment.

Overpressure can also occur in liquid lines completely filled with liquid and sealed (with valves) at each end. As temperatures fluctuate, enormous pressures can result from the expansions involved.

The first rule to reduce hazards from such hydrostatic overpressures is to remove all unnecessary valves at the design stage. Rarely does any line require a valve at each end, and any valve not really needed for operation will seize up and no longer function anyway. Remove all unnecessary valves. If a line has to be fitted with two valves, the space between should be be pressure-relieved, or the two valves should have an interlock to prevent both being simultaneously closed.

8.3c RELIEF EQUIPMENT DESIGN

The equipment must be specified with overpressure relief equipment, and for two grounds: condenser failure and fire. Should the condenser fail, then the boiler will continue to heat the column, and vapour will continue to be produced. The pressure will rise, the liquid boiling point will rise, until the design pressure is reached, at which point the relief system will operate. The capacity of the relief system must be specified so that it handles the quantity of vapour produced. This involves knowing how much vapour and knowing the pressure driving force across the relief system.

The vapour quantity can be determined from the heat transfer across the boiler to the boiler contents at the new elevated boiling point due to the increased pressures. The driving force to dispel this quality of vapour is the design pressure of the equipment less the back pressure in the vapour handling system. Thus we have all the factors necessary to design the pressure relief system, and the size of the equipment - be it an overpressure relief valve or bursting disk - can be calculated.

Design of relief in a fire situation is achieved taking known radiation heat transfer rates and evaluating the corresponding vapour rate from submerged surfaces. This will generally be less than the boiler vaporisation rate for a distillation column and thus will not be the design condition for the relief equipment. Relief equipment on receivers and stock tanks which can be isolated, however, have to be designed to cope with the fire situation.

8.3d DESIGN TEMPERATURE AND PRESSURE

The design temperature and pressure have to be defined by the chemical engineer responsible for specifying the distillation equipment. The vessel designer is then able to design the vessel to withstand the defined design temperature and pressure.

The design temperature is the highest temperature that the equipment is likely to achieve in the event of a failure. This could safely be taken as the temperature of the heating medium. If this brings serious consequences - the need for expensive materials of construction for instance - consideration could be given to protecting (by additional instrumentation) against the eventuality causing such a high temperature, so that a lower design temperature can be specified.

The design pressure is the highest pressure that the equipment is likely to attain in abnormal, but conceivable operation. The designer uses the design pressure with strength of materials at the design temperature to determine the vessel thickness to use. The equipment will then be safely operable under these conditions, but any higher pressure must be relieved by the pressure relief system. Hence the choice of design pressure is again an optimization. A low design pressure will keep costs lower but may cause operating difficulties if the relief system opens frequently. In many cases, for example in equipment with pressure less than 5 bar, little is saved by specifying low design pressures, since the equipment must support its own weight, and this usually results in a vessel that will support 5 bar pressure.

Distillation columns must withstand considerable vacuum, since this can easily develop in an enclosed system if the boiler steam is cut off and sufficient inerts are not fed into the column. Inflammable systems should not be fitted with vacuum relief systems enabling air to enter in events of a vacuum developing.

The chemical engineer should be involved in setting the corrosion allowance.

This is the thickness of the material allowed to be removed from the surface
before the thickness equals that required for design temperature and pressure
considerations.

When corrosion rates are known from handbooks, an economic choice of materials of construction and equipment lifetime can be determined. The designer
must decide whether it is economically best to chose a material of construction
that will last 15 years or an initially cheaper solution of a less expensive
material of construction with a corrosion allowance that will make it last for
5 or 10 years, and then allow for the cost of replacing it after this time.
Such a study requires a discounted economic analysis to properly account for
the different timing of the costs before the optimal choice can be made.

The chemical engineer and the materials scientist should together define the
corrosion conditions, select the materials of construction and find, or measure, the corrosion rate, so that a corrosion allowance can be specified.

The final wall thickness is taken as that calculated from design pressure considerations plus the total corrosion allowance - made up from corrosion rate
from both the inside and the outside of the wall.

8.4 FURTHER READING

Proc. Workshop - Industrial Process Control, Florida, 1979. A.I.Ch.E.
- Some excellent survey papers by Luyben, Buckley and Mosley, describing
   industrial distillation practice.

Rademaker, O., Rijnsdorp, J.E. and Maarleveld,A., 1975. Dynamics and Control of
   Continuous Distillation Units, Elsevier.
- 700 pages on distillation control; very complete, but detail hides the main
   principles.

Shinskey, F. G., 1984. Distillation Control : for Productivity and Energy Conservation, McGraw-Hill.
- good, particularly for complex distillation systems.

8.5 QUESTIONS TO CHAPTER 8

1) What are the two basic requirements that must be fulfilled for a column to
   be properly controlled?

2) What problems can arise in using temperature to indicate composition in the
   column?

3) Why is no one standard control system suitable for all distillations?

4) What advantages are sought when cascade and feed-forward systems are
   specified?

5) Why does distillation equipment require relief systems?

6) What is the design pressure and design temperature of a piece of equipment?
   What factors are brought into consideration before they are specified?

Chapter 9

PILOT EXPERIMENTS AND DEBOTTLENECKING

9.1 PILOT TESTS FOR NEW PROCESSES

Whenever new processes are developed, be they new chemical processes or new distillation processes, then it is important that adequate design data be obtained and pilot trials made before very large sums of money are committed for a full scale plant.

The first requirement in the case of distillation is to have available the VLE data for the key components. This is obtained from the literature, predictions and ones own measurements.

If the separation has been measured and reported in the literature over the concentration range of interest then it will be a great start to the study. Looking for measured VLE data from the literature used to require a painful search through chemical Abstracts, but over the years a number of data collections have been published which makes the work easier. Hala first published VLE data collections, but this has now been superseded by the Gmehlin and Onken DECHEMA Data publications, which is a thorough collection and analysis of most measured VLE data.

If a literature search shows that the VLE has not been measured, the prediction of the VLE data should be attempted. The most complete prediction method available is the UNIFAC method, as described in Chapter 2. This method predicts VLE from the interaction of the various groups in the molecules involved. It therefore follows that it cannot be used unless the appropriate group interaction parameters have been obtained from other systems. It often happens in mixtures occurring in chemical processes that some interaction parameters are missing and so UNIFAC can then not be applied.

The accuracy of the UNIFAC method for VLE prediction of completely new mixtures is difficult to judge because all measured data has been brought into the data base. It is surprisingly accurate, but to predict difficult separations and azeotropes sometimes requires extremely accurate data. Therefore the UNIFAC authors' words of caution should be heeded: The UNIFAC method can indicate whether there is likely to be a distillation problem. If the method shows the distillation is not simple, experimental measurements must be made.

If our system is not reported in the literature, and the UNIFAC method suggests the distillation will need more than a few plates, experimental measurements will be necessary.

Within a company the normal way is to write a request to the Physical Chemistry and Analysis Group to carry out a VLE study on the new system. The chemical engineer must then wait some weeks for the results, which charges his project some thousands of dollars for the laboratory study.

9.1a LABORATORY MEASUREMENTS OF VLE

In general, VLE of binary systems are measured in the laboratory and fitted to multicomponent VLE models such as WILSON or UNIFAC, so that multicomponent behaviour can be predicted from the measured binary data.

Laboratory measurement of MULTICOMPONENT systems is to be avoided because of the vast amount of work it involves, since it is a combinatorial problem. Binary studies are generally possible because of the confidence in the use of binary pair coefficients to predict multicomponent behaviour. However, if the predicted multicomponent separation does look difficult, then the laboratory measurement of some multicomponent points to determine the accuracy of the predicted points would be advisable. If this shows that the prediction of multicomponent data for binary coefficients was not accurate enough, then multicomponent data will have to be measured in the laboratory. Onken and Gmehlin contained some measured data on multicomponent systems, which at least can be used to quickly get an indication of the accuracy of using binary data for multicomponent prediction.

The range of concentration over which the VLE is measured should cover the concentration range over which distillation in the final processes must operate. The majority of VLE reported in the literature contain 9 equidistant points between 0.1 and 0.9 mole fraction. This is most unsatisfactory if the column is to produce high purity products, since the majority of the column will be working in concentration ranges outside the 0.1 to 0.9 mole fraction - concentrations where no laboratory measurements were made, concentrations where maximum deviation from ideality can be expected, and where extrapolations from the 0.1 to the 0.9 measurements can be expected to give maximum error.

THE VLE EQUILIBRIUM STILL

VLE data is measured in the laboratory by allowing vapour to come to equilibrium with boiling liquid. At equilibrium, samples of liquid and vapour are taken and analyzed, temperatures and pressures are measured, and so the data is captured.

There are numerous problems. To allow equilibrium to be achieved requires a form of recycle system to be devised that will maintain the quantity of liquid phase constant. The same vapour must be recycled which involves condensation and revaporisation, both to supply a liquid sample for analysis and to obtain the pressure increase necessary to achieve circulation. It is also not a simple matter to obtain a vapour in equilibrium with a boiling liquid; boiling is a rate process and so none-equilibrium conditions exist.

The requirement for the apparatus is that the separation be exactly one theoretical plate. In a recycling system as all concentrations become steady, there is little problem in liquid mixing (or vapour mixing), but there are problems with excessive wetted contact areas. If the walls above the liquid are wet, then further mass transfer occurs with a liquid film that is not of the same composition as the liquid on the plate. Hence further vapour enrichment occur as the apparatus corresponds to more than one theoretical plate. Wetted walls with a counter-current liquid film on any surface before the condenser have the same effect, and so the equilibrium still design must be careful to eliminate this situation.

Numerous types of apparatus have been devised for overcoming these problems, and they are fully reported in the literature. Figure 9.1 shows one such apparatus - the Othmer still.

For normal and low pressure systems, normal glass apparatus is used. For high pressure systems, then stainless steel equipment has to be designed, which, although very different in appearance, is based on the same principles.

VLE measurement must be carried out with very pure components. Any impurities could collect preferentially in the liquid or vapour phases, or could affect the analysis or the total system pressure. Dissolved gases can be removed by a boiling and purging operation in the still before the run proper is begun.

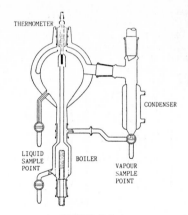

FIGURE 9.1
THE OTHMER VLE EQUILIBRIUM STILL

It is clear that the request to have VLE measurements carried out involve many
weeks of work. Raw material must be obtained and purified, analytical procedures
set-up, and the still degassed and run to equilibrium for each point measured.
The results then carefully analyzed, correlated by regression and finally reported.

THE RESULTS

The results of the laboratory VLE study are the VLE data for the key components. This is generally not sufficient to enable the final design to be made
and the full scale equipment to be ordered with confidence because:

(a) Only the key component VLE data has been supplied. What about the other
components? Where will they appear in the column products? Will they interfere
with the separation?

(b) Belonging to the natural world, there are a host of other effects besides
the separation that are important for the design of a distillation column. Are
the boiler contents stable, or are there thermal decomposition or coking problems? Will there be corrosion problems from acids formed from boiler pyrolysis
or water enrichment in the column? Does the mixture present foaming problems?
Are there dangers of blocking, plugging, or encrustation?

Unless the system in question is a clean system with a limited number of well
known components, then it is likely that a pilot distillation trial must be
carried out before enough confidence has been gained to go ahead on the full
scale.

9.1b THE PILOT DISTILLATION

If it is decided that the distillation stage of a new process should be
piloted, than the first question is "how big".

Pilot distillations can be carried out in the laboratory glass column of 50 mm
diameter, in small conventional columns of say 100 mm diameter, or even in
small industrial columns of 500 mm diameter. Questions of course are; How much
material is available? What pilot equipment is available? What information do I
loose if I work at too small a scale?

LABORATORY OLDERSHAW COLUMNS

Oldershaw columns (Figure 9.2) are glass sieve plate columns 25 or 50 mm diameter, with a downcomer arrangement giving about 3 mm of liquid on each plate.

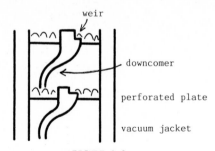

FIGURE 9.2
THE OLDERSHAW COLUMN

It is a miniature sieve plate, as near as the craftsmanship of the glassblower will allow. The miniaturisation of the liquid height means that the resulting froth in no way is resembles the froth on an industrial sieve plate, and the low limiting gas velocities of Oldershaw columns means that different surface tensions effects are observable in Oldershaw columns (the Marangoni Effect) which play no role in full scale columns.

The Oldershaw plate is not a pilot sieve tray, but an Oldershaw column is a multistage distillation with known reflux and known tray efficiency of about 0.6. The Oldershaw COLUMN is therefore an excellent pilot column, since the full scale will be a multistage distillation, with reflux, with a stage efficiency of about 0.6.

Oldershaw piloting of distillation is an established practice in the chemical industry. Standard laboratories are erected with a number of Oldershaw distillation units available. If a new separation has to be investigated, the liquid mixture is brought to this distillation laboratory, an Oldershaw system constructed (Figure 9.3) and the separation tested. When satisfactory conditions have been obtained, the corresponding number of plates, feed point and reflux are simply transposed to the full scale design.

In one experiment the separation of key components has been checked, the fate of minor components known, and any foaming, decomposition and corrosion observed.

The disadvantage however is that the achievement of a satisfactory distillation configuration has been the result of trial and error - each new reflux demanding a new run, each new feed point a column modification, and each new number of plates a complete column rebuild.

The work takes many months, and a satisfactory configuration can be delivered, but no VLE data; i.e. no information to enable re-thinking during the design or improvement in performance of the plant when built, is available as a result of the laboratory study.

SMALL SCALE CONVENTIONAL COLUMNS

If a new process is being piloted, there is uncertainty over the distillation stage, and product is required for marketing evaluation, then it is sensible to include a distillation column as part of the pilot plant and use this for the pilot distillation work. Such a column would be a "small conventional column".

The column internal should be those considered most suitable for the final scale. Packing sizes would have to be reduced to maintain good liquid distribution with the 10:1 diameter ratio rule. Plate columns can be installed in small diameter columns by manufacturing cartridges containing a number of plates which are lowered into the column.

The difference between the pilot column and the small manufacturing column is in the facilities provided for measuring what is happening in the column. Sample points must be provided for top and bottom products and also for column profile measurement down the column. Flow rate must be accurately measurable to allow heat and mass balances over the column to be determined to give credibility to the experimental data.

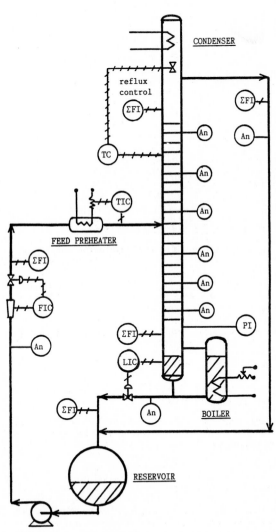

FIGURE 9.3
OLDERSHAW PILOT DISTILLATION TEST SYSTEM

Of utmost importance in the removal of "internal reflux" in the column, caused by heat loss and progressive condensation up the column, leading to flow conditions that cannot be reproduced on the large scale column. "Internal Reflux" is the major problem in using small columns for distillation testing. In Oldershaw columns the problem is reduced by having vacuum jacketed, silvered sections. On a small scale pilot plant it may be advisable to install electric strip heating in the lagging around the column, and control it so that the lagging strips are at the same temperature as the inside of the column.

It is advisable to be able to measure the reflux at the top and bottom of the column, to check whether the measures taken to remove internal reflux have been successful. The top reflux ratio is easily measured (as long as a deflegmator is not used as a partial condenser). The liquid flow rate into the boiler is rather more difficult to measure unless the liquid flows out of the column, or can be collected into a bucket device within the column and the time observed to fill the bucket.

The major disadvantage with this equipment for experimental work is that the configuration of the full scale plant is fixed by the design of the pilot plant. There is no possibility to try a different number of plates, or different internals. This equipment will confirm that the design will function, but the designer must be fairly certain of what he needs before the pilot plant is built.

SMALL INDUSTRIAL COLUMNS

Many companies have general pilot facilities, and these require the preparation or recovery of various solvents or chemicals. For this routine work it is useful to have a reasonably sized column, and once this is available it is ideal for use for pilot trials.

Such a column can be of completely conventional design, of 200 to 500 mm diameter, fitted with boiler, condenser and fully instrumented. Since it is to be used for obtaining data, further analysis points, to enable component profiles to be measured, and better feed measurement for heat and material balances, should be provided.

Accuracies that can be achieved with such equipment are astounding. If product receivers and feed tanks can be weighed (with an overhead weight-scale on a travelling crane), mass and heat balances within 1% can be achieved, which is a very good foundation from which to start to interpreting the results.

The main disadvantage with such a column is that its number of plates cannot easily be changed. The only variables are feed point position and reflux ratio. Hence, the effect of number of plates is rather difficult to determine. A second disadvantage is that it requires considerable quantities of feedstock.

9.1c ANALYSIS OF OPERATING COLUMNS

We have discussed three types of pilot distillation equipment. All involve many months of work, and some are very expensive. How can we interpret the results from these experiments?

The Oldershaw column delivers a number of plates, feed point and reflux that will achieve the required separation. It is not usual to relate these results to any VLE measurements, since this only confuses the otherwise simple picture. No attempts can be made when designing the full scale equipment to consider alternative designs to reduce costs, improve flexibility etc., as demanded in the earlier chapters. The design point is that achieved by an experimental search in the laboratory, usually made within very strict time constraints.

The small scale conventional column will deliver the required reflux ratio and
reflux position to best provide the required separation with a given number of
plates. It confirms (or disproves) the design made by the pilot plant designer,
and so, if successful, allows the designer to use the same design method for
the design and optimization of the full scale plant. If the results do not confirm the design, or if the pilot design was "specified" without the use of a
design method or VLE data, then this equipment also does not contribute to the
full-scale designer's objective of investigating different possible designs
before specifying the best.

The Small Industrial Column again delivers results that confirm or disprove
the designer's information. If VLE data is available, plate efficiencies can be
calculated, and these use to indicate whether the system is understood - as
will be discussed later.

If reasonable plate efficiency data is obtained, the designer is confident that
he can design this or any other configuration, and he can consider different
designs for this full scale. If the results show he cannot predict the performance, than there is little more the results tell him.

In all these cases, months of work are expanded, and there is no new quantitative data produced, despite the fact that many measurements have been taken and
much numerical experimental data obtained. Because of this very unsatisfactory
situation, a project within the Swiss Federal Institute (ETH) was begun and has
developed the following approach for extracting more information from the pilot
plant work.

THE HYKA METHOD

The principle of this method is to measure concentration profiles from experimental pilot plant distillations, and use non-linear regression techniques to
determine the VLE data that must have given rise to these profiles. Once VLE
parameters have been obtained, they can be used of the designer in the normal
way for designing full scale equipment (Rose, 1984).

The idea is no different to the analysis of chemical reactors where parameter
fitting of profiles is used to determine the reaction kinetic parameters. It is
however, far from obvious that such a procedure can be successful. Computer
times may be excessive, parameter fit unsatisfactory, and correlation between
parameters rendering parameter values worthless. Only by extensive tests and
program development could it be shown that such a proposal can be used to ex-
tract useful design information from pilot plant results.

The multicomponent continuous distillation model relates the distillate and
bottoms compositions to the operating variables such as material and heat input flows and compositions,and reflux ratio. Further variables define equipment
details such as number of plates and feed plate position. Required also is the
VLE model relating vapour composition to liquid composition, which itself requires a set of VLE parameters, and the plate efficiency for the plate and system in question.

The analysis of an operating distillation column requires the determination of
plate efficiency and appropriate VLE parameters (for a given column from a number of sets of distillate and bottoms compositions at given operating conditions) which enables the model to fit the experimental data.

Additionally, intermediate liquid concentrations can be measured and so the top
and bottom product compositions can be augmented to give a larger set of intermediate liquid composition data which can also be used in the fitting of the
parameters.

Since the information contains experimental error, sufficient data has to be measured to overdetermine the system so that the best fit can be obtained by least-squares fitting. Since the equations are non-linear we are faced with a non-linear regression problem.

The most obvious way to pose the regression problem is to let the operating conditions be the independent variables, and the concentration profile (and the product compositions) be the dependent variables, with the overall plate efficiency and the VLE parameters being the parameters to be determined by the regression.

For multicomponent systems this requires iteration of the distillation model to achieve convergence of the equations, and this means the model will take a considerable time to compute. Since this will be repeated some hundreds of times by the search procedure of the non-linear regression, it can be expected that such a proposal will introduce unacceptably high computer costs.

If we depart from the above proposal, and assume that the bottom product mol fractions are the independent variables, then the computational requirement of our model will be drastically reduced. Once the bottom composition is known, and in this case it has been experimentally determined, the we can proceed with a plate-to-plate calculation from the bottom of the column to the top and the column calculation need only be done once.

The advantage of this calculation procedure is two-fold. Firstly there is a reduction of 5-20 times in the computational time, and secondly the non-convergence problem associated with difficult distillations is circumvented.

In principle, therefore, in the HYKA method a small number of distillation runs are made; operating conditions and concentration profiles are measured; and the profiles fitted by means of a non-linear regression package, using the measured boiler compositions as the independent variables in a simple plate-to-plate once-through distillation model. The regression delivers the values of the VLE model parameters and overall plate efficiency which gives the best fit to the experimental profiles.

The development of this procedure, however, showed that for it to be reliable, a number of other factors have to be carefully considered. These are handled as individual topics in the following paragraphs.

Model Stability

The regression search moves the VLE parameters about extensively some hundreds of times in a computer run. The distillation model has to yield a result under all circumstances. This requires recovery routines to cover all eventualities of mol fractions being outside the range 0-1, and care is needed with exponentials which could become out of range as well as log and square root statements. In these non-real cases the accuracy is irrelevent as the resulting profile is anyway very poor and the objective function being minimised correspondingly large.

Weighting

Since the concentration in the fitted profile varies from a few ppm to 100%, each response has to be appropriately weighted to prevent low concentrations being ignored and only high concentrations being relevant to the fit.

It addition to this weighting to correspond to experimental error, a second weighting factor was sometimes necessary to 'steer' the regression in a physically real region.

VLE Models

The HYKA method was developed with the Wilson VLE model. The following discussion is therefore only relavent to the Wilson model, although the UNIQUAC model could equally well have been used.

The Wilson equation has two parameters WC_{ij} and WC_{ji} for every binary pair in the system. Hence, the number of parameters to fit easily becomes large, and this introduces problems in the regression. The number of parameters can be reduced by the following analysis.

The two Wilson parameters can be defined by the interaction energies between the different molecules λ_{ij} and interaction between similar molecules λ_{ii} and λ_{jj}.

$$WC_{ij} = \lambda_{ij} - \lambda_{ii} \tag{9.1}$$

$$WC_{ji} = \lambda_{ij} - \lambda_{jj} \tag{9.2}$$

λ_{ii} and λ_{jj} are equivalent to the negative of the energy of vaporisation which can be related to the heat of vaporisation (ΔH) by:

$$-\lambda_{ii} = \Delta H_i - RT \tag{9.3}$$

where ΔH, according to the Clausius-Clapeyron equation can be related to vapour pressure:

$$\frac{d(\ell n P_i)}{dT} = \frac{\Delta H_i}{RT^2} \tag{9.4}$$

Using the 3-term Antoine equation to describe vapour pressure the λ_{ii} term can therefore be related to the Antoine constant as follows:

$$-\lambda_{ii} = RT^2 \frac{B_i}{(C_i + T)^2} - RT \tag{9.5}$$

where B and C are the Antoine constants for component i.

Hence, λ_{ii} and λ_{jj} can be calculated from pure component vapour pressures and so it is possible for a reasonably ideal system to reduce the Wilson model to one parameter - λ_{ii} -per binary pair. This known as the 1-parameter Wilson model and it fits VLE data extremely well in many cases. The advantage of the 1-parameter Wilson model in this work is the reduction in the number of parameters to be fitted.

The 1-parameter Wilson model in many cases gives excellent fits and fast location of the parameters. With the 1-parameter model it is possible to include the overall plate efficiency as a parameter for the search, resulting in 1 more parameter to be fitted. There are, however, some cases where the 1-parameter model is not accurate enough, and a 2-parameter model is necessary to obtain the required accuracy.

For some systems even the 2-parameter Wilson equation cannot represent the system accurately enough and we have a problem of the introduction of systematic error from an inadequate VLE model. The demands put on the VLE model in the analysis of operating columns are extreme. Relative volatilities are below 2.0, otherwise the distillation would not be a problem anyway, with the operating line near the equilibrium line to achieve low energy costs. Hence a 5% error in the prediction of the activity coefficient can represent a 10 to 50% error in the separation per plate.

Plate Efficiency

Consideration of plate efficiency is a major discussion point in this method. In _theory_, plate efficiency is dependent on plate design, physical properties and the slope of the equilibrium line. In _practice_, for difficult distillations, the plate efficiency is generally close to 70 ±10%. Hence, the assumption that the plate efficiency is constant over the whole column is not unreasonable in practice.

In many cases the mean plate efficiency can be identified in the analysis by either entering it as a parameter to be fitted by the regression or by locating the value that gives the minimum sum of squares in a sensitivity study. In systems where this is not possible because of correlation between unknown parameters, a value must be set which is typical for the equipment in question.

When the analysis can identify the plate efficiency, the the HYKA method offers a further advantage. Although the fitted plate efficiency cannot be use for a scaled-up column directly, if the system has shown a suprisingly different plate efficiency to that expected on the test column, then one knows to design for an equally unusual efficiency on the full scale.

Starting Parameter Values

The response surface for the parameter search is very complex and possesses more than one minimum. Hence, randomly chosen initial values for the Wilson parameters have a low probability of resulting in a successful run. The starting values were therefore calculated as those Wilson parameters which gave an ideal system (i.e. activity coefficients as unity at infinite dilution, which is a function of the component liquid densities).

Experimental Region

There are considerable advantages in carrying out the experimental work at total reflux: flows do not have to be measured, and concentration ranges are widest.

The disadvantages with total reflux work can be listed as:

- the top product can be unrepresentative if impurities or degradation products collect at the column head,

- the column is operated at the furthest point away from industrial operation with regards to reflux ratio,

- Adjusting the concentration range of the experiment is more difficult.

The concentration range of the experimental work must be equal or greater than that to be used on the full scale. This is the major criterion in the choice of experimental conditions. Two experimental concentration ranges are useful to ensure that the total region has been covered.

Fitting Procedure

The fitting was carried out during the development of the method with the Marquardt algorithm which was specially modified to improve its robustness. It cannot be claimed that the fitting of the parameters is easy, but, given an understanding of the distillation process, a quick turn-around at the computer and a systematic step-wise fitting procedure using the weighting facilities, then there are no unsurmountable difficulties.

Typical computer times for a parameter fit for two profiles for a ternary system with 25 plates is about 100 seconds on a CDC 6500.

Some Experimental Results

The method so described was shown to satisfactorily predict VLE information
from profile data for 6 systems, on 3 scales of equipment; a 50mm Oldershaw
column, a 300mm sieve plate column and a 2000mm sieve plate column.

Figure 9.4 indicates the closeness of fit that is obtained by the fitting procedure. Figure 9.5 shows 3 examples of the plate efficiency determination made
by a sensitivity study to determine the plate efficiency that gave the minimum
sum of squares. In cases (a) and (b) it is convincing that the plate efficiency
is 0.56, but in case (c) no conclusion can be drawn. This is because the plate
efficiency is correlated with the Wilson parameters in this last case. This may
be because there was insufficient experimental data, or it could be inherent in
the system of equations.

The HYKA method therefore offers a method for analyzing pilot plant data – if
the effort is made to obtain concentration profile data and measure all flows
accurately – to produce VLE parameters which are then usable at the design
stage to fine-tune the reflux ratio, number of plates, feed position, etc..

The method "fills a void" as the academic say. It is a pity that its publica-
tion in the respectable literature was delayed seven years by the adverse reaction of the VLE experts. Refereeing serves little purpose but to ensure that
papers to be published are very similar to papers that have been published.

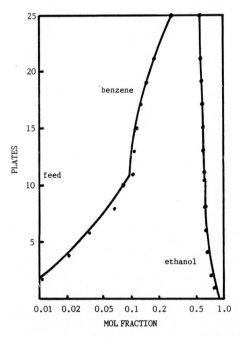

FIGURE 9.4
FITTED EXPERIMENTAL PROFILE BY THE HYKA METHOD

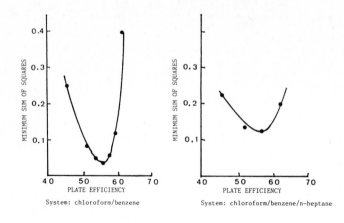

System: chloroform/benzene   System: chloroform/benzene/n-heptane

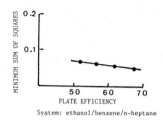

System: ethanol/benzene/n-heptane

FIGURE 9.5
PLATE EFFICIENCY DETERMINATION WITH THE HYKA METHOD

9.1d SCALE-UP AND PLATE EFFICIENCY

We can therefore obtain pilot plant results, analyze them if necessary, but the full scale plant will have a different plate design - how do we handle the scale-up of the plate efficiency?

Chapter 7 handles plate efficiency in some detail because for the column dimensioning the understanding of plate efficiency becomes a necessity.

Our full scale plate will be properly designed - it will not weep or entrain excessively; the liquid height will provide adequate mass transfer. It will no doubt have an efficiency of about 0.7. If the column is large it may have a higher efficiency because of the enrichment due to crossflow and incomplete liquid mixing.

If tests have been made on the plate we are going to recommend, than we could use the theory of Chapter 7 to predict the change we will expect in plate efficiency due to our particular system, with its new m, D_L, D_V, and L/V. However we know these corrections will be minor.

The test with the pilot column may have included plate efficiency measurement for this column, or the efficiency could have been determined by running a test mixture through the column. Again it is likely to be between 0.5 and 0.7 - possibly a little less than the full scale. Hence the number of plates required for the full scale can be scaled from the pilot column by the ratio of the plate efficiencies.

If we include a 10% safety factor on the full scale, the number of plates for full scale will be very close to the number in the test column. This confirms the rule of thumb of more than 25 years standing: "The number of plates in the full scale column equals the number of plates in the Oldershaw test". Physically the two columns are completely different, but it is easy to see why this rule holds. The method has, rather belatedly, reached the scientific literature (Fair,1983).

If plate efficiency is straightforward, why does it cause so much concern? In fact, plate efficiency is very useful to measure because it is so straightforward. If VLE data is available and the test column has operated, then the plate efficiency can be calculated. This should be very similar to measurement on a column with a test system (and definitely between 0.5 and 0.9). When this is not the case, then there is something wrong:

- Is the VLE accurate? False assumptions of ideality or the use of data fitted at too high concentrations leads to apparent errors in plate efficiency.

- Is the plate correctly installed? Incorrect plate operation, wrong weir heights and flooding produce incorrect efficiencies. The same arguments apply to packed columns when using HETP instead of plate efficiency. High HETP's are produced by incorrect liquid distribution (in one case because a sack containing the packing had accidentally fallen into the column).

- Is the incorrect plate efficiency due to basic errors in our understanding? Are reactions involved that were not anticipated?

The plate efficiency (or HETP measurement) is a useful indicator to see that the system is sufficiently well understood to be designed full scale. It should be measured and discussed, but not treated as some fudge-factor not to be questioned.

The follow 3 case studies may provide relevant entertainment:

CASE STUDY 1 - PURIFICATION OF CARBON TETRACHLORIDE

A new process for the production of carbon tetrachloride was being developed by the further chlorination of chloroform by the free-radical reaction with chlorine at 300 ° C. The carbon tetrachloride so produced had to be purified for use in Freon 12 manufacture, which demanded very low limits of other chlorinated hydrocarbons in the carbon tetrachloride.

The reaction stage was successfully demonstrated, and about 10 tons of crude product was collected. Although there was considered to be no problem in designing distillation equipment to attain the required specification – since all halogenated hydrocarbons behave fairly ideally – is was decided a full scale distillation trial on plant size equipment should be made. A first run through the column removed lights and chloroform, and a second run through was to separate heavies from the carbon tetrachloride product.

The quantity of trichlorethylene was not reduced as expected by this second distillation. To investigate further, a careful test of the separation at <u>total reflux</u> in the industrial size column was made. The column was a packed column with liquid redistribution, and this gave the opportunity to collect a number of liquid samples and measure the concentration profile in the column. From this profile the HETP between each sample point could be calculated. The result showed the HETP in the top and middle of the column were normal, but at the bottom of the column the HETP increased many-fold. The single total reflux run with its measured profile delivered very useful information: - The separation was difficult at the bottom of the column at low concentrations.

As with many industrial projects, the work stopped before the final explanation was found. One theory being that the system was not "ideal" as the analysis as-

sumed and that non-idealities showed themselves at the low concentrations. The
second theory being that high chlorinated ethanes decompose slowly at the boiler temperature giving a steady production of trichlorethylene which gave the
apparent poor separation.

The moral of this story is it is useful to measure HETP (or plate efficiencies)
even at total reflux where no flow measurements are required because it is a
good indicator of trouble.

CASE STUDY 2 - **ABSORPTION** OF HF IN AQUEOUS HCL

A process was being developed where gaseous HCl containing HF was stripped of
HF by counter-current washing with aqueous HCl. The HF had to be removed to
prevent dissolution of the plant downstream which was made of glass.

HF solubility data in aqueous HCl was available and the absorber confidently
designed. The resulting **absorption** plant did not work and the HF proceeded
in the destruction of the glass plant. Absorption experiments with a single
plate (Wolff bottle) showed that absorption efficiencies were very low, but
given time (days) concentration of HF in the aqueous HCl rose considerably. The
problem seemed to be in the absorption efficiency or rate,not in the solubility
data.

Very low efficiencies are indicative of slow chemical reactions - somewhere a
reaction was involved in the system. Discussion with the analyst showed that
the reported HF content of the HCl gas stream - by tradition - was always the
HF after complete hydrolysis. The major component was carbonyl fluoride which
was sparingly soluble in water but which hydrolyzed slowly to HF.

Faces were red all round, the absorption plant was shut down and the thinned
glass plant replaced.

Again it was an efficiency measurement which indicated that something was not
properly understood.

CASE STUDY 3 - MEASUREMENT OF INDIVIDUAL FILM RESISTANCES

Around 1955 - 1960 it was fashionable to carry out tray performance studies
using air/water systems, and then relate them to a mixed film distillation
performance using the two film theory, as discussed in Chapter 7 .

$$\ln (1 - E) = \frac{1}{k_g a} + \frac{mV}{L} \frac{1}{k_L a} \qquad\qquad (9.6)$$

Vaporisation of a solvent is pure gas film controlling, and desorption of a
sparingly soluble gas is pure liquid film controlling. Hence measurement of the
vaporisation of the water from a plate operating with an air/water system, and
measurement of the stripping of oxygen from an oxygen enriched liquid feed ena-
bled $k_g a$ and $k_L a$ to be evaluated, and the prediction of mixed film systems such
as those to be expected in distillation to be predicted.

Experiments were carried out on the type of equipment traditionally used for
such studies - see Figure 9.6.

Single film measurements were carried out, and used to predict desorption of
methanol, acetone and ammonia - 3 mixed film systems - from water with air.

The prediction failed. For example, from efficiencies of 0.81 and 0.92 for water and oxygen plate efficiencies, a prediction of 0.70 for mixed desorptions

was compared to the experimental measurement of 0.56. The experimentally determined plate efficiencies were all lower than that predicted by the combination
of the individual films. Something was clearly wrong.

Looking again at Figure 9.6, 3 regions have been marked; A, B and C. The area
of the froth could be calculated from other work and it was clear that A presented as much as 50% of the interfacial area of the actual froth B. Similarly
the wetted wall of the empty running downcomer presented 20% of the froth area.

Now looking at the effect of A and C on the mass transfer on the plate; for water evaporation , A is a wetted film, just as effective as B, whereas area C
will not contribute because there is very little gas flow. The opposite applies
to oxygen stripping; the liquid film on A will be stripped of oxygen and contribute nothing to the mass transfer, but the area C will be active as far as
oxygen desorption is concerned.

For mixed film controlling systems only B will be effective; for pure gas film
controlling A + B; and for pure liquid controlling B + C. No wonder the
predictions were poor.

Adequate predictions could be made from one mixed system to another mixed system, or using the pure film results after "correcting" for these additional
areas A + C. However the concept of characterizing small plates with pure film
resistance systems is basically false – even though it was a fairly wide-spread
procedure for university studies.

Again, it was the inability to explain the plate efficiency measurement that
pointed out the presence of a problem.

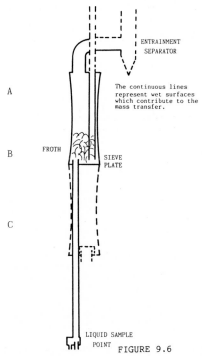

FIGURE 9.6

SINGLE SIEVE PLATE FOR AIR/WATER SINGLE FILM STUDIES

9.2 PROCESS DEBOTTLENECKING STUDIES

If product sales are going well, and more output is required, then the question arises as to how much can the plant ultimately produce? Unfortunately, all that is usually known about the ultimate plant capacity is the ultimate capacity of the bottleneck in the process - it may be a flowmeter, valve, pump, heat exchanger or reactor! The real questions are:-

- what should the new rating of the plant be?

- how many items should be expanded?

A prediction of the ultimate performance of all items is required to answer these questions, and these must be predictions based on performance at sub-ultimate capacities. Rarely can an item be tested to its limiting capacity because of the limitations of other units.

The distillation section is often one of the more significant parts of a plant, and whether it is operating near its maximum capacity or not can be decisive in deciding how far to debottleneck the plant.

The prediction of the ultimate capacity of a distillation system is an interesting study. It should be based upon reliable measurements and not correlations because at the debottlenecking stage errors of 10-20% are very significant, whereas for design these are tolerable. Often VLE is incompletely known and plate efficiencies, pressure drop, flooding characteristics are all required to be known with more accuracy than would be expected for a new design.

The degree of overdesign in the equipment can be tested by cutting back on the reflux until the top and bottom products become out of specification. This should be followed up by a feed point investigation to see whether the feed point is ideally positioned for the new (or even the original) reflux ratio.

The next step is to measure the separation achieved (preferably measuring a full composition profile) and relate this to the plate efficiency (or HETP) if VLE data is available. Such tests are best done at a high reflux ratio (or even at total reflux) as long as the products are still quantitatively analysable, since with restricted reflux,large sections of the column do not enter into the separation because of the mass balance equations. Such "dead sections" of a column play havoc with the accuracy of the plate efficiency measurement. With the plate efficiency measured, then a calculation study can be made to choose the best conditions to maximize production for the existing column.

If VLE data is not available, we can still measure our profile and use the HYKA method to determine what the VLE data and plate efficiency must have been. This data can then be used to complete a calculation study to determine the optimum conditions as though VLE data had been available.

Prediction of the ultimate hydrodynamic capability of the column should be based on pressure drop measurements over the column, and a comparison with published or manufacturers curves. If the operating point reveals a displacement of the pressure drop versus Loading Factor curve published for the column internals in question, then this displacement can be assumed to operate over the whole range, and the displacement in maximum loading determined.

Short term hydrodynamic maximum load tests will be useful if the plant can make off-specification material for a short while when rates are increased temporarily, and if boiler and condenser capacity are adequate. Maximum heating, increased reflux ratio and lower pressures are always of means of producing higher loadings, and there may be one or more of these possibilities open in any particular case.

CASE STUDY 4 - PLANT DEBOTTLENECKING

Product demand required that a plant capacity be substantially increased. The key item in the plant was a 50 plate 2 m dia. extractive distillation column, of unknown ultimate capacity. Should this capacity be more than 150% of the rated capacity, extensive plant debottlenecking sufficed. If it was less, a new plant must be installed. The column was originally specified on the basis of pilot trials, the available VLE data was inadequate for design purposes. Predictions with the UNIFAC method were shown to be too inaccurate for such a debottlenecking study.

Feed and product compositions, pressure and reflux ratios were variables available to a total process optimization study, but without an accurate quantitative model of the distillation nothing could be investigated.

Four runs were carried out with the industrial column within the limitations imposed by the capacity of the other items of the plant by making small changes in reflux ratio and operating pressure. Product samples were taken as were a limited number, namely 4, vapour samples from the column. The HYKA method was used to identify the VLE parameters from 2 of these runs and these same parameters used to predict the remaining two runs.

The prediction of these runs gave a measure of confidence for the other predictions from the model. Hence, a quantitative model was available to enable the debottlenecking study to start.

It should be emphasized that experimental conditions were far from ideal in this work - a very restricted number of sample points were available (mainly pressure tappings in the column) and the column control system was far from steady, resulting in an unsteady concentration profile. Nevertheless VLE data was obtained and an helpful column model was developed.

This model was then used within a flowsheet program to thoroughly investigate the different operating possibilities for the whole process. The flowsheet program available for this study did not have a non-ideal multicomponent distillation model available, so that a short-cut Fenske-Gilliland model was used for the flowsheet study.

This required that the effect of reflux ratio and pressure on the top and bottom specifications be predictable by the Fenske model. Simulations were carried out with the full model, using parameters derived using the HYKA method, and then relative volatilities for the 3 components were found which gave effectively the same performance as the full model over the limited reflux ratio and feed composition ranges of interest. This reduced model was then used in the flowsheet for the total process investigation.

9.3 COLUMN UPRATING

When the debottlenecking study shows that the column ultimate capacity is inadequate for the output required from the operated plant, then it is necessary to consider uprating the column itself.

Many equipment items can be uprated by adding further units - heat exchanges or compressors for instance. Columns, however, are normally such large pieces of equipment that replacement is out of the question and addition of a parallel unit also has its problems.

Columns are such large costly items that if their output cannot be increased, this often represents the ultimate capacity of that plant. Conversely, if the output can be raised by 30% this makes a significant saving in not having to build a new plant.

9.3a CHANGE OF INTERNALS

Raising the ultimate capacity of a column is concerned with changing its internals for a more modern and more suitable design. Consideration can also be given to changing the reflux ratio to move along the reflux ratio/number of plates curve to find a point at which a greater output can be achieved, accepting that the internals or product specification can also be changed.

PACKED COLUMNS

It is clear from looking at the performance data that Pall rings have a better capacity per cross-section column area than Raschig rings. A column operating with Raschig rings (specified perhaps many years ago, or specified originally for low cost) can therefore be re-packed with Pall rings and the column will have been uprated. Metal Pall rings will give a further uprating.

It may be that a slightly smaller packing with a better HETP (e.g. some form of saddle) may allow sufficient reduction in the reflux ratio to overcome the lower vapour capacity and still result in an uprating of the quantity of feed that can be handled. There are a wide range of packings available, and the choice of a more modern packing together with a small change in its size may provide the best uprating.

There is opportunity to specify packings in the uprating study that the original designer did not have. The original design would have been based on minimum total cost, with due consideration for plant spares policy. At the time of the uprating study, the cost picture has completely changed. If a new packing can be found to uprate by say 30%, then the cost of a new plant is saved. With this economic picture the more sophisticated packings – including the structured packings – become economic if they will provide the uprating.

For an uprating study, detailed performance of the various packings must be compared and this requires performance data which is best obtained from the packing manufacturer or from internal company tests. The operation of columns is so important a subject that usually internal company studies have defined policy as to the best packing for uprating studies, and as time progresses, the company accumulates more and more knowledge and operating experience with this recommended packing.

PLATES

As good fortune has it, most old plate columns were fitted with bubble cap trays. Since almost all trays are better in capacity and efficiency than bubble cap trays, this has provided many years of profitable work, particularly in the oil industry, in re-traying the old bubble cap columns with better performance trays.

Sieve trays are a better proposal than bubble caps, but valve trays may be the better choice if turn-down is a problem. The lost area in plate columns due to the downcomers is appreciable, and to insert trays without downcomers gives an immediate capacity increase because of the greater available free area for the vapour. Turbogrid trays are therefore good for uprating existing columns. Their operating range is very narrow, but, for a production expanding situation, it may not be necessary to discuss turn-down.

The amount of money saved can be enormous. The re-traying of a pipe-still may be a first step to the uprating of a refinery by 30%. Usually large petroleum and contracting companies have made careful studies of these possibilities and have decided on a policy – as with the packed columns – which becomes a standard procedure for re-traying company-wide.

It is the re-traying of columns that causes there be so many different types of tray available. It is hardly worth the original designer going beyond standard sieve trays or valve trays in his original design, but for the re-traying exercise, trays with minor advantages at large extra costs can become interesting propositions. In fact, the original designer would not be thanked for being too clever because there would be no possibility for later uprating these columns where demand warranted it.

9.3b PRESSURE MODIFICATION

Column vapour loading is normally correlated by the loading factor F, where F is $v\sqrt{\rho_g}$. As the density (i.e. pressure) of the gas doubles, then the column capacity increases by the square root of the increase - 40%. For an atmospheric column there is little problem in increasing its pressure to 5 bars, because it is generally strong enough to stand such operating pressures. This must however be properly checked before decisions are taken.

The problem, however, lies in the separation. As a general rule the higher the pressure, the lower the relative volatility and the more difficult the separation. This phenomena can be useful in breaking azeotropes, but in the case of uprating columns it is often so serious - particularly as the column is designed near minimum reflux anyway - that rarely is the pressure increase a practical solution to column uprating.

The combination of pressure uprating and smaller packing to give more plates may be attractive, but such a policy is not common in practice.

9.3c ADDITIONAL COLUMNS

When all else fails, then additional columns have to be provided to uprate the distillation capacity. In theory this is best done by installing a second column operating identically as the first column in parallel.

In practice there is usually some existing column on the site with the wrong number of plates and the wrong diameter which is available at little cost if only it can be incorporated in some way to increase the distillation capacity.

Such a column could be used before or after the existing column to pre-concentrate the feed, or work-up the top or bottom products to be within specification, enabling the specification at one end of the existing column to be relaxed to obtain the increased capacity. Figure 9.7 shows some possibilities. No rules exist for such situations. The best solution is found to each problem by a specific case study.

Distillation design column programs are an important tool in such a study, and a flowsheet program incorporating good distillation models would be even better as a means of locating the best solution.

9.4 ENERGY CONSERVATION

It almost looks as though the days of debottlenecking and uprating are coming to an end, and present work with existing processes is aimed at reducing the energy demand for the process. Distillation is a large energy consumer and so there is great potential for energy saving. This is particularly true of older plants where the energy recovery schemes were right when the energy costs were 1/20 of the present energy costs.

Clearly all old plants can be studied and their energy household improved because the objective functions have changed so much.

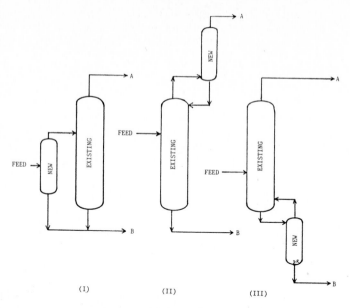

FIGURE 9.7

VARIOUS WAYS OF INCREASING COLUMN CAPACITY BY THE INCORPORATION OF
A FURTHER COLUMN

The traditional energy recovery systems are concerned with feed pre-heat and
product cooling (in pipestills for example). Clearly new temperature levels and
new exchanging networks are going to be more economic than the older designs,
and the amount of money involved is usually sufficient to justify major re-
vamps in the energy recovery scheme. Details of how this can be done is discussed a little in Chapter 6, but the design of heat exchange networks is not
really within the scope of this book.

More involved is the use of heat of condensation from one column for boil-up
in a second column operating at lower temperatures. Again vast sums of money
are involved, and so incentives are large for coupling columns in this way.
From an operational point of view such a coupling is detrimental to stability,
but this can be overcome if supplementary heating is available so that
fluctuations in one column can be compensated for by controlling the
supplementary heating to the second column.

Consideration should be given to the modification of the system pressures to
achieve suitable temperature levels for these heat exchange systems to be employed. It must, however, be remembered that pressure changes effect the relative volatility, and require higher utility temperature levels, so that there
are many constraints restricting the scope for changing pressures in trains of
distillation columns.

Rather than attempting to change pressures in different columns to achieve
compatable temperatures, it may be more sensible to use condensers as reboilers
part-way up other columns, where temperatures are lower (see Figure 9.8). This
changes the L/V ratio at the base of the column, which reduces the separation,
but in many instances the critical column section for separation is around the
feed, and the L/V is less important at the ends of the columns. The use of a
multicomponent distillation program with facility for sidestreams can be use to
checking the effect of the intermediate boiler on the separation to see if a
fraction of the total boiler heat load can be added higher up the column before
bottom product specification becomes a problem.

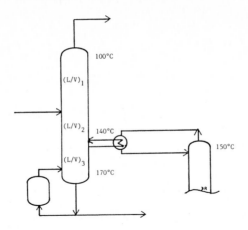

FIGURE 9.8
SUPPLYING BOIL-UP HEAT PART-WAY UP THE COLUMN

Distillation columns are ideal candidates for heat pumps; heat loads are equal and temperature levels are not too far apart (particularly for difficult separations). Admittedly electricity is at present so much more expensive than steam (particularly low pressure steam) that heat pumps are difficult to justify economically, but one can expect them to become more economically competitive, and easier to install as improvements to existing plants in the future.

Particularly suitable would seem to be the incorporation of a heat pump for a part of the vapour load at either side of the feed point, where L/V ratios are critical. This gives a small difference in temperature levels, a good Coefficient of Performance for the heat pump, and leaves normal equipment to deal with the top and bottom of the column. This, like the previous scheme, changes L/V ratios top and bottom which have a detrimental effect on separation, but this may be compensated for by changes in reflux ratio or in packing type in these parts of the column.

It can be expected that most existing columns will be re-examined to determine the optimum reflux ratio to use in view of the changing energy costs. The tendency will be towards lower reflux rates and more plates, which for existing columns, means repacking with packing with improved HETP, or re-traying with more efficient trays, to run at the lower reflux ratios.

The change in energy costs may have its disadvantage, but for the chemical engineer it provides many opportunities to exercise his creativity to improve existing plants.

9.5 FURTHER READING

Fair, J.R., Null, H.R. and Bolles W.L., 1983. Ind. Eng. Chem. - Pro. Des. and
            Dev.,22, p53.
               - using laboratory columns to predict full scale.

Gmehlin, J. and Onken,U., 1977. Vapour-Liquid Equilibrium Data Collection,
            DECHEMA, Frankfurt.
               - Extensive collection of VLE data.

Rose, L.M. and Hyka,J., 1984. Ind. Eng. Chem. - Pro. Des. and Dev., 23, p429.
               - analysis of operating column distillation data.

9.6 QUESTIONS TO CHAPTER 9

1) What methods are available for obtaining VLE for the design of distillation columns?

2) What types of pilot distillation are used, and how can their results be interpreted?

3) What information can be obtained from 'plate efficiency' measurements on pilot columns?

4) Why are particularly accurate prediction methods necessary for debottlenecking studies?

5) When there are advantages in changing column internals, does it mean that the designer originally made a poor choice?

6) What methods for energy conservation can be applied to existing distillation plant?

APPENDIX 1

SOME USEFUL CONVERSIONS TO SI UNITS

	SI unit	<u>Divide</u> by this factor to convert to SI units
Length	1 m	= 3.281 ft = 39.372 inches
Area	1 m^2	= 10.765 ft^2 = 1550 inches^2
Volume	1 m^3	= 35.31 ft^3 = 220.0 imperial gallons = 264.2 US gallons = 6.290 barrels
Mass	1 kg	= 2.205 lbs = 1×10^{-3} tonnes = 0.9842×10^{-3} long tons
Temperature	K	T°C = (T + 273.15) K $T°F = \dfrac{(T - 32.0)}{1.8} + 273.15$ K
Pressure	1 bar	= 0.1×10^6 kg m^{-1} s^{-2} (Pascal, N m^{-2}) = 0.9870 atm = 14.51 psi = 750.1 mmHg = 401.5 inches water
Diffusivity	1 m^2 s^{-1}	= 1×10^4 cm^2 s^{-1}
Surface tension	kg s^{-2} (N m^{-1})	= 1000 dynes cm^{-1}
Density	1 kg m^{-3}	= 0.001 g cm^{-3} = 0.06243 lb ft^{-3} = 0.01002 imperial gallons = 0.008345 US gallons
Energy	1 kJ (N m)	= 1 kW s = 1 kg m^2 s^{-2} = 0.2778×10^{-3} kW h = 0.239 kcal = 0.9478 BTU = 737.6 ft lbs
Power	1 kW	= 1 kJ s^{-1} = 1 N m s^{-1} = 1 kg m^2 s^{-3} = 0.001 amp x volt = 1.340 HP = 860 kcal h^{-1} = 3412 BTU h^{-1}
Force	1 N	= 1 kg m s^{-2} = 0.1×10^6 dynes = 0.1020 kilopond = 0.2248 lb force
Volumetric flowrate	1 m^3 s^{-1}	= 3600 m^3 h^{-1} = 2119 ft^3 min^{-1} (cfm) = 13200 imperial gallons min^{-1} (gpm) = 543.6×10^3 barrels day^{-1}

	SI unit	Divide by this factor to convert to SI units
Henry's gas solubility constant	1 bar m^3 kmol^{-1}	$= 0.987$ atm (mol l^{-1})$^{-1}$ $= 750.1$ mmHg (mol l^{-1})$^{-1}$
Heat-transfer coefficient	1 kW m^{-2} K^{-1}	$= 860$ kcal m^{-2} h^{-1} °C^{-1} $= 176$ BTU h^{-1} ft^{-2} °F^{-1} $= 79.9$ kcal h^{-1}ft^{-2}°C^{-1} $= 0.02391$ cal cm^{-2} s^{-1} °C^{-1}
Thermal conductivity	1 kW m^{-2} K^{-1}	$= 2.391$ cal cm^{-1} s^{-1} °C^{-1} $= 557.7$ BTU h^{-1} ft^{-2} (°F ft^{-1})
Heat capacity	1 kJ kg^{-1} K^{-1}	$= 0.239$ cal g^{-1} °C^{-1} $= 0.239$ BTU lb^{-1} °F^{-1}
Gas mass-transfer coefficient	1 kmol m^{-2} bar^{-1} s^{-1}	$= 3647$ kmol m^{-2} atm^{-1} h^{-1} $= 0.1013$ mol cm^{-2} atm^{-1} s^{-1} $= 747.0$ lb mol ft^{-2} atm^{-1} h^{-1}
Liquid-phase mass-transfer coefficient	1 m s^{-1}	
Viscosity (dynamic)	1 kg m^{-1} s^{-1} (N s m^{-2})	$= 10$ poise $= 1000$ centipoise $= 10$ g cm^{-1} s^{-1} $= 0.672$ lb ft^{-1} s^{-1}
Velocity	1 m s^{-1}	$= 3.281$ ft s^{-1}

APPENDIX 2

SOME PHYSICAL PROPERTY DATA

It is important for a designer to appreciate the magnitude of the relevant physical property data, and also the magnitude of the variations to be expected between different components. The following table gives some data for some compounds commonly met in distillation. The units are SI.

Component	Mol. Wt	Bpt.	Sp. Heat*		Density*	Viscosity*		Surface Tension*
		K	kJ/kmol K		kg/m^3	Ns/m^2 x 10^{-4}		N/m
			Liq	Vap	Liq	Liq	Vap	
Benzene	78.1	353.3	154	105	791	2.55	0.0990	0.0182
Butane	58.1	272.7	170	118	465	1.13	0.0967	0.0040
CCl$_4$	153.8	349.9	150	90	1454	3.90	0.1334	0.0166
Chloroform	119.4	334.9	128	72	1270	2.86	0.1280	0.0171
Ethanol	46.1	351.4	157	85	721	3.62	0.1063	0.0243
n-Heptane	100.2	371.6	262	204	610	1.96	0.0792	0.0120
Methanol	32.0	337.8	155	58	703	2.28	0.0960	0.0255
Methylcyclohexane	98.2	374.1	224	174	639	2.96	0.0864	0.0155
m-Xylene	106.2	412.3	219	159	798	2.83	0.0841	0.0204
1-CH$_3$-Naphthalene	142.2	517.8	249	209	957	8.47	0.0735	-
P-xylene	106.2	411.5	220	158	791	2.86	0.0837	0.0200
Toluene	92.1	383.8	187	131	782	2.64	0.0918	0.0199
Water	18.0	373.1	76.4	34.2	957	2.76	0.1265	0.0610

* Properties evaluated at a temperature of 373 K.

APPENDIX 3

CAD EXERCISES

The following 20 exercises have been devised for solution using the computer programs presented in Appendix 4. They are all extensive, and are suitable as mini-projects of one or two half-day periods duration. It is best if students work together in small teams of 2 to 4 per team, and have one terminal available per two teams. A printer should be available in the terminal room to which selected outputs can be send. It is also possible to run the exercises with batch computer facilities, as long as a quick turn-round (e.g. 15 minutes) is achievable. Punched card facilities can even be used; these at least allow for easy modification of data without have to master editing or file handling syntax.

Experience has show it is advisable to give the students a period of 30 to 60 minutes to study the problem before they are allowed on the computer. This period enables them to read the computer manual, and do some simple hand calculations to get a feel for the problem before going on the machine.

vilvegnuege!

EXERCISE 1 - USE OF A DATA BANK

Use the data bank to determine the following:

a) Molecular weights and boiling point for ethanol and water

b) Pure vapour pressure of ethanol and water at $25°C$ and at their boiling points.

c) Pure liquid viscosities, surface tension and liquid density at $70°C$ of tetrachloroethylene.

d) The temperature dependence of the surface tension of water by obtaining its value at $-50°C$, $25°C$, $100°C$ and $200°C$.

EXERCISE 2 - DATA BANK FOR ENTHALPY CALCULATIONS

Use the databank to determine:

a) The enthalpy of 1 kmol of pure methanol under the following conditions:
   - $25°C$ vapour
   - $25°C$ liquid
   - boiling point vapour
   - boiling point liquid

Compare the results with the Heats of Formation and the Latent Heat. What is the basis for the enthalpy figures determined by the bank, and why has this base been choosen?

b) 274 kmol of a mixture of benzene and heptane (0.4 mol fraction benzene) at $40°C$ is to be heated and vaporised at 2 bar pressure. What is the total heat load to be supplied by the vaporiser?

c) A plant produces 1 ton/hour of carbon tetrachloride by reacting chloroform with chlorine (100% conversion).

$$CHCl_3 + Cl_2 \rightarrow CCl_4 + HCl$$

The reactor feed is at $100°C$ (gaseous) and the exit $320°C$. How much heat has to be removed from the reactor?

EXERCISE 3 - DATA BANK FOR FLASH CALCULATIONS

Use the databank to determine:

a) The boiling point of a mixture of ethanol and water (0.1 mol fraction ethanol) at 1 bar pressure (compare ideal, Wilson and UNIFAC treatments)

b) Can the above mixture be separated by a single flash into relatively pure components? - What vapour and liquid compositions would be achieved?

c) How many plates would be needed to obtain a 0.5 mole fraction ethanol vapour product at total reflux from the mixture given under a)? (This is equivalent to repeated boiling point and dew point calculations).

d) What is the composition of the ethanol/water azeotrope as predicted by the UNIFAC method? What is the azeotropic composition assuming ideality?

EXERCISE 4 - THE UNIFAC METHOD

a) From the information given in the CHEMCO manual determine the group numbers, and numbers of each group which compose the molecules;

   - propylene
   - benzene
   - methanol
   - chloroform
   - chlorobromomethane

Compare your answers with the CHEMCO stored groups (Method 19) where possible.

b) Can the UNIFAC method be used to calculate VLE from a mixture of;

   - propylene/benzene
   - chloroform/chlorobromomethane

c) How many different ways can the molecule methyloxyisopropanol:

$$CH_3 - O - CH_2 - CH - OH$$
$$\quad\quad\quad\quad\quad\quad\quad\quad | $$
$$\quad\quad\quad\quad\quad\quad\quad CH_3$$

be described using UNIFAC groups?

EXERCISE 5 - BINARY DISTILLATION

A column is required to recover methanol from a mixture of 0.40 mf methanol and
0.60 water. The recovered methanol should have a purity of 95 mol percent, and
the discharged water should contain less than 1 mol% methanol. The feed rate is
400 kmol/hour.

Data: Antoine Coefficients

	A	B	C
(1) Methanol	11.8550	3571.47	-36.507
(2) Water	11.8684	3927.82	-41.813

Vapour/Liquid Equilibrium

	WC_{12}	WC_{21}
Wilson coefficients	861.0	2030.0

Liquid Molar Volumes

	A	B	C
(1) Methanol	0.0405	0.0	0.0
(2) Water	0.018	0.0	0.0

A plate column is to be used, with a plate efficiency of 0.70. The design F
factor vapour loading in the column can be considered to be 2.0 $kg^{1/2}_m{}^{-1/2}s^{-1}$. Under these conditions the plate pressure drop will be 0.01 bar/plate. Propose a
suitable column design and choose an optimum reflux ratio.

For cost data the following approximate data can be taken:

        Steam cost   10 $/tn $_2$
        Investment cost 3000 $/m^2 of plate

EXERCISE 6 - BINARY DISTILLATION (2)

Heavy water (D_2O) is to be recovered from waste water returned from heavy water
user plants.

The waste water contains 20 mol% D_2O, and the D_2O recovered must have a purity
greater than 95 mol%. The plant capacity is to be 2 kg/hour of D_2O.

Determine the number of theoretical plates and reflux ratio required for the
separation. Neglect pressure drop for this initial analysis. What pressure
should be used for the operation?

Data: Antoine Coefficients

	A	B	C
H_2O	10.8741	3383.74	-62.13
D_2O	12.4034	4212.73	-33.6

Ideality can be assumed for the VLE calculation.

Design to 1.4 times the minimum reflux ratio.

Use FENGIL or BINARY program or both as appropriate.

EXERCISE 7 - MULTICOMPONENT DISTILLATION (1)

A column is required to separate 100 kmol/h of the mixture

Mol Fraction

Benzene	0.15
n-Heptane	0.35
Toluene	0.40
m-Xylene	0.10

into an n-heptane fraction and a toluene fraction. The n-heptane fraction should contain less than 0.01 mol fraction toluene, and the toluene fraction should contain less than 0.02 mol fraction n-heptane.

Determine the number of plates and reflux ratio to achieve the separation. Assume the internals can be selected to have a pressure drop of 0.3 mbar/theoretical plate.

Begin with a Fenske program run to obtain the initial estimate and a suitable reflux ratio, and then use the program DISTIL to detail the design.

Data:

Antoine Coefficients

Component	A	B	C
Heptane	0.92533E+01	0.29128E+04	-0.56351E+02
Benzene	0.92850E+01	0.28091E+04	-0.50295E+02
Toluene	0.94613E+01	0.31344E+04	-0.52059E+02
m-Xylene	0.92380E+01	0.31850E+04	-0.67000E+02

Use UNIFAC for the prediction of VLE data

- what is the best feed plate?

- what are the differences between the Fenske result and the DISTIL result?

- is this difference due to VLE model assumptions?

EXERCISE 8 - MULTICOMPONENT DISTILLATION (II)

A column is required to produce Chloroform of adequate quality for Freon and PTFE production. It must therefore meet a specification of 0.1000 wt% CCl_4 and 0.005 wt% CH_2ClBr. 90% recovery of Chloroform as product is required.

The column feed is 70 kmol/hour with a temperature of 40°C and concentrations of 0.20 mf CCl_4 and 0.002 mf $CH_2Cl Br$. With a plate efficiency of 0.70 and a design F loading of $1.8 \, kg^{1/2} m^{-1/2} s^{-1}$ design a distillation column and specify the control system. Plate pressure drop at design load is 12 mbar.

Data

Antoine Coefficients

Chloroform	12.4932	4833.91	52.7951
Chlorobromomethane	12.4682	4857.08	48.8367
Carbon Tetrachloride	12.2540	4866.59	47.9317

The system can be considered to be ideal.

	Liq. density	Mol. weight
Chloroform	1490 kg/kmol	118.5
Chlorobromomethane	1600 kg/kmol	129.5
Carbon Tetrachloride	1590 kg/kmol	154

Propose a design of distillation column. In particular:

- what is reasonable reflux ratio?
  (start with a Fenske run)

- how many plates are required?

- where is the best feed point?

EXERCISE 9 - BATCH DISTILLATION

Design a batch distillation plant to handle 8 m^3/batch of Benzene, Heptane, light oil mixture (mole fraction 0.6, 0.3, 0.1) with a batch cycle time of 12 hours.

The product purities required are:

- Benzene 0.95 mf
- Heptane 0.95 mf
- Oil to contain less than 10 mol % lights

(a) Determine a suitable size of column, boiler and condenser.

(b) What should the intermediate fraction reflux ratios be?

(c) Should the boiler be supplied with additional area as internal coils.

Data:

Benzene	mol wt	78.11		
	Antoine	12.717	5017.34	41.8
	mol vol	0.0975		
	L-enthalpy	156		
	V-enthalpy	31400		
Heptane	mol wt	100.21		
	Antoine	11.997	4719.91	22.5
	mol vol	0.0125		
	L-enthalpy	200		
	V-enthalpy	40200		
Oil	Mol wt	200		
	Antoine	11.100	4720.00	10.1
	mol vol	0.25		
	L-enthalpy	400		
	V-enthalpy	80000		

The system can be considered to be ideal.

Overall heat transfer for both boiler and condenser can be assumed to be 2 kW/m^{2}K. Hold-up should be neglected.

EXERCISE 10 - BATCH DISTILLATION (2)

A plant is required to recover heavy water from 2 grades of waste
water returned from heavy-water system containing 20% and 80% heavy water respectively. 3 tons are available with 20% D_2O and 2 tons with 80% D_2O. 90.% purity D_2O is required in the final product.

For such low quantities, batch distillation should be used. Propose different
distillation schemes for achieving the separation. Use the BATCH program to
evaluate the most promising scheme, and determine the dimensions of the equipment required.

Data:

Antoine Coefficients	A	B	C
H_2O	10.8741	3383.74	-62.13
D_2O	12.4034	4212.73	-33.6

Ideality can be assumed. Hence it is possible to use the relative volatility
short-cut option in the BATCH program.

EXERCISE 11 - BATCH DISTILLATION WITH HOLD-UP (1)

The batch equipment design in Exercise 9 is to be checked for the effect of
hold-up in the column. Make reasonable assumptions concerning equipment volumes
to determine the condenser hold-up, and consider two cases for the column hold-
up.

(a) a sieve plate column is to be used with a plate efficiency of 0.7 and a
clear liquid level on the plate of 30 mm.

(b) Sulzer structured packing is to be used with a 6 volume % hold-up at
working conditions.

What is the influence of the 2 internals on the distillation performance?

Assume the feed mixture for the distillation contains only 0.08 mf heptane with
0.8 mf benzene and 0.12 mf light oil. What is the effect of the 2 internals in
this case?

EXERCISE 12 - BATCH DISTILLATION WITH HOLD-UP (2)

Taking hold-up into account, choose suitable column internals for the batch
heavy water distillation equipment requested in Exercise 10.

Use hold-up data given in the BATCH manual (Appendix 4).

This will require repeated simulations with the BATCH program with different
hold-up quantities.

EXERCISE 13 - TEMPERATURE CONTROL POINT

Consider the temperature profile for Exercise 5 and select a suitable point for
the temperature control.

How much will this temperature move when the column is running at 50% of design
rate?

Consider the possibility that the column may have a feed of 60 mol % methanol and 40 mol % water. What feed rate could the column which was designed for the Exercise 5 condition handle, and where should the new feed and temperature control points be located?

Specify the column in terms of number of plates, feed point and temperature sensing point which would give a suitable robust flexible design.

Either the DISTIL or BINARY programs could be used to carry out the investigation.

EXERCISE 14 - COLUMN SEQUENCING

The stripped product from a methane chlorination contains either (a) or (b), which has to be separated by distillation

	kmol/h	
	(a)	(b)
Methyl Chloride	1.0	20.0
Methylene Chloride	2.0	15.0
Chloroform	15.0	2.0
Carbon Tetrachloride	20.0	1.0
Total Feed	38.0	38.0

Use heuristic rules and the program FENGIL to determine good sequences of distillation columns to produce the four products, each with a specification of 0.98 mf, for the above two feeds. What are the opportunities for heat recovery.

Data:

	Antoine Constants			Latent Heat
	A	B	C	kJ/kmol
Methyl Chloride	10.6480	2697.74	4.19	21,600
Methylene Chloride	11.5036	3597.39	0.0	28,000
Chloroform	12.4932	4833.91	52.79	29,600
Carbon Tetrachloride	12.2540	4866.59	47.93	30,100

EXERCISE 15 - AZEOTROPIC DISTILLATION

Ethanol, from 90 kmol/h of a 25 mol % aqueous ethanol mixture has to be recovered with 99 mol % purity. 95% of the ethanol fed should be recovered.

This will require the azeotrope to be broken, and this will require a 2 column system using either benzene as the ternary component, or using a two pressure system.

(a) Use the data bank to determine the composition of the normal azeotrope and to determine the effect of pressure on composition.

(b Use the data bank to determine the potential for breaking the azeotrope using benzene.

(c) From rough compositions, determine a flowsheet and mass balance for the two methods.

(d) Use appropriate distillation program software to investigate these two systems in detail.

(e) Revise the mass balance and produce a final process design.

Data:

All the data should be taken from the data bank. The UNIFAC VLE model must be used. Although the system with benzene may form 2 liquid phases in the column and thus form two phases in the condenser, still useful information can be obtained assuming the phases do not separate using a 1 liquid phase analysis.

EXERCISE 16 - PLATE DESIGN

Dimension the column for methanol/water separation investigated in Exercise 5. A turn-down of 30% is required by the production people.

A plate column is to be chosen. Select a suitable plate design and column diameter, taking into account entrainment and downcomer flooding, froth height and plate efficiency. Check the proposed design at different points in the column and determine whether the required turndown can be achieved with a sieve plate.

The INTERN program can be use to check the performance of the proposed plate.

EXERCISE 17 - COMPARISON OF INTERNALS

Dimension the column for heavy water separation already investigated in Exercise 6. Consider different column internals (sieve, Pall rings and Sulzer packing) and use typical specific pressure drops (PD/theoretical plate) to determine the effect of pressure drop on the separation.

Dimension the column for each of the packing types, and compare the results in capital costs.

Specific pressure drop data is tabulated in Chapter 7. Use the BINARY program with pressure drop to determine the number of plates, and the INTERN program to size the column. Cost data correlations are also given in Chapter 7.

EXERCISE 18 - CONTROL SYSTEMS

Leaving the reactor in the Freon 12 process is a mixture of:

	mol %	Bpt at 5 bar $^\circ C$
HCl	45	-50
Freon 12	45	+16
HF	10	+80

This stream enters the first column in order to separate HCl from the other two components. There is a specification on the HCl that it contains less than 0.1 mol % HF, and for economic reasons the loss of Freon 12 with this HCl should be minimized. The bottom specification is very loose since it will be water washed and HCl within this stream simply represents the financial loss of HCl which anyway has a very low value.

The column is controlled at a pressure of 5 bar, and the coolant is boiling refrigerant at -50°C. This is very expensive and it is essential therefore that the condenser heat load be minimized. The HCl top product is to be taken off as a gas.

Draw the reactor/column/condenser/boiler part of the process, propose a total control concept and draw in the resulting control system.

EXERCISE 19 - COLUMN UP-RATING(1)

A column for the separation of Benzene and Toluene from a 40/60 mol % feed mixture was laid out rather generously and has always produced material much better than the specification.

The plant is to be debottlenecked, and it is necessary to know what the column capacity will be when the column produces material which is just in specification. Furthermore, the bottom specification can be relaxed to allow for a further capacity increase.

The present column is old and was fitted with some patent internal whose data has been lost. How much can the plant be uprated without modifying the column?

Present working conditions:

	Operating column			Original specification		Relaxed specification	
	Feed	Top	Bottom	Top	Bottom	Top	Bottom
Flows kmol/h	80	32	48				
mol % benzene	40	99.9	0.2	99.0	1.0	99.0	5.0
mol % toluene	60	0.1	99.8	1.0	99.0	1.0	95.0

Reflux ratio 3.6
Feed point   60% of the column height
Loading   70% flooding

What conditions can be changed to get an increased throughput, and by how much?

Data:
Antoine Coefficients

Component	A	B	C
Benzene	0.92850E+01	0.28091E+04	-0.50295E+02
Toluene	0.94613E+01	0.31344E+04	-0.52059E+02

Use the UNIFAC VLE model if available, otherwise use ideality.(Firstly simulate the operation of the existing column to determine the number of plates in the column).

EXERCISE 20 - UPRATING (2)

A column for the separation of a 50% mixture of two naphthalene derivatives requiring uprating.

The present column operates at a bottom pressure of 0.2 bar, with a reflux of 4:1 and has 20 old bubble cap trays with an efficiency of 0.60 in the column of height (for internals) of 9.0 m, and 1 m diameter. The pressure drop is 0.10 bar.

(a) What increase in capacity can be expected if other internals are used, simply in terms of increased loading of the 1 m diameter column?

(b) What further benefits are obtained if allowance is taken of the reduced pressure drop given by the new internals? (Boiler temperature is critical for determining column pressure, but column flooding occurs at the lowest pressure point).

(c) What further improvement in throughput can be made if use is made of the increased number of theoretical plates given by the better packings.

Assume ideality, and calculate the effects using the FENSKE program. The system has a relative volatility of 1.75.

Programs FENSKE and INTERN are useful for the solution of this exercise.

APPENDIX 4

COMPUTER PROGRAM USER MANUALS

This Appendix contains four user manuals for the programs mentioned in the text, and recommended for the solution of the CAD exercises. Four sets of manuals are presented:

(1) CHEMCO data bank manual

(2) DISTILSET distillation programs manual containing BINARY, FENGIL and DISTIL user instructions.

(3) BATCH batch distillation user manual

(4) INTERN column internals user manual

These programs have been specifically designed for teaching at undergraduate level and are available at nominal costs from EURECHA (The European Committee for Computers in Chemical Engineering)

        EURECHA
        c/o SEG-group,
        Technisches Chemisches Labor,
        ETH-Zentrum,
        CH-8092 Zurich,
        Switzerland.

In addition to the 4 programs sets detailed in this Appendix, other programs with themes related to distillation are also available from EURECHA:

- THERMDINSET   Equations of state, fugacities, UNIFAC with a 168
                 component data bank.

- VLESET   Programs for correlating laboratory VLE binary
                 measurements with the WILSON, NRTL and UNIQUAC
                 models. Programs for determining the accuracy
                 of ternary data predicted from binary coefficients
                 against measured data.

- CAPCOS   A computerized version of the complete Guthrie
                 capital cost estimation procedure. Capable of
                 delivering cost estimates for individual items
                 up to total plants, inclusive of off-sites.

- DIAGNOSE   A trouble-shooting exercise using plants with
                 simulated faults which have to be diagnosed by the
                 student. A distillation plant is included.

- SYNSET   Two synthesis programs for the synthesis of sequences
                 of distillation columns and the synthesis of
                 heat exchanger networks.

E U R E C H A

The European Committee for the Use of Computers in
Chemical Engineering Education

"CHEMCO" PHYSICAL PROPERTIES DATA BANK
M A N U A L

(Program Version July 1984)

Manual Author: L.M. Rose
                    Technisch-Chemisches Labor
                    E.T.H. Zentrum
                    CH-8092 Zürich
                    SWITZERLAND.

CONTENTS   Page

1. INTRODUCTION

The EURECHA physical properties data bank CHEMCO has been specifically prepared
for teaching purposes in that it demonstrates most of the features of data banks
but does not have the completeness or guaranteed accuracy of a commercial bank.

This in turn means that its storage requirement, running cost and user description are all much smaller than for an industrial bank.

The storage of data is on a sequential file which, although not necessary for the comparatively small data set involved, at least demonstrates the principles and difficulties of sequential storage. This requires that there are always two calls necessary to the bank: firstly to announce that a property will be needed and secondly to evaluate that property. The first call enables all the requirements to be collected together and all the required parameters to be collected from the archive and stored in local commons in a single search through the archive. Besides demonstrating these principles it also is the most efficient form for use in engineering programs where multiple calls for the same property are made.

The physical properties covered by the bank are a minimum set needed for chemical engineering design work;

    MWt, BPt, MPt, Heats and free energy of formation
    heat capacities
    Enthalpies
    Thermal conductivities
    Viscosities
    Molar volumes
    Vapour pressure
    Surface tension
    Wilson and UNIFAC VLE models.

Apart from UNIFAC none of the methods are predictive but are based on fits to experimental data. Hence, the bank is suitable for all types of compound, inorganic, organic and hydrocarbon, but it does not adequately demonstrate the use of the equations of state and physical property prediction methods.

Mixture properties are delivered by the bank using simple molar averaging methods. A comprehensive FLASH package is part of the bank and this delivers results for the following 6 types of flash calculation:

    Liquid vapour pressure
    Boiling point
    Dew point
    Partial vapourisation
    Isothermal flash
    General adiabatic flash

Temperature, pressure and liquid and vapour compositions are given for each type of calculation and the VLE model used can be Ideal, Wilson or UNIFAC. All mixture and flash calculations are limited to a maximum of 10 components in the mixture.

The archive supplied with CHEMCO contains 63 common compounds which have been found useful in student design projects and distillation studies. Permanent addition of further compounds is straightforward and can easily be done for instance before a particular student design project is begun. The existing archive contains about 20,000 data items, i.e. quite small by normal data bank standards.

Provision is made by means of the COMPAD subroutine to temporarily add new components or missing data to the local common storage to enable mixture and flash calculations to be carried out for specific studies without having to extend the archive with rarely used or unreliable data. Data added with COMPAD is not stored after the run has finished.

The archive has to be generated with a program GEN which takes parameter values for properties and compounds in a random order, sorts them and produces the ordered archive file which has to be loaded whenever CHEMCO is being used. The generation procedure also ensures consistency between liquid and vapour enthalpies and the heats of vaporisation.

The bank can be used in two distinct ways. By means of an interrogative program GETDAT the bank can simply provide physical property data --- in the form of parameter values to the equations, point values for pure components or mixture values. GETDAT is an interactive program but it can be used with a batch system by correctly anticipating the questions that will arise. COMPAD is available to GETDAT so that mixture values can be provided even when all the components are not on the archive.

The second use of the bank is with engineering programs where the engineering program calls the CHEMCO subroutine which then delivers point values or parameter values for the required physical properties. Flash calculations can similarly be called to deliver liquid and vapour compositions. COMPAD is available when all components or parameters are not on the archive.

CHEMCO has been used for many years for teaching both with GETDAT to provide point values and with the flowsheet program UNICORN and various distillation programs. In its old form - without UNIFAC - it has been distributed to many universities.

The CHEMCO data bank is the result of co-operation between a number of European universities. Milan provided the basic coding from their hydrocarbon bank PHYSCO. Liège provided the parameters for the archive generated from their bank EPIC. Lyngby provided the UNIFAC models and data from their UNIDIST program. Zürich co-ordinated the project and coded the necessary interfaces and finally the Hungarian Institute for Science Management and Informatics modified the input coding to make it flexible and free-read.

The FORTRAN used has been strictly ANSI 66 (except for data statements) and names kept within the limits of 16 bit machines. The program is therefore extremely easily transferred, provided adequate storage is available. The total set of programs contains about 2000 statements and requires a storage of the order of 20 K.

No guarantee of accuracy of the data provided by the bank can be given. Though based on accurate EPIC data, the use of simplified models, the neglecting of non-idealities and the omission of any final check on the accuracy of the results means that the bank should not be used for industrial purposes without firstly checking the accuracy of the data delivered by the bank.

2.   THE DATA BANK SYSTEM

All the data is stored on a file referred to in this manual as the Archive. The basis of the system is the subroutine CHEMCO, which extracts the required data from the archive and holds the data in a common area where it can be used by the various model subroutines to provide point and mixture data, where it can be retrieved by engineering programs for use in engineering and flowsheet calculations, or where it can be used by the FLASH subroutines.

Following principles of good file handling the CHEMCO routine firstly assembles all the data requrests and then searches the file only once to retrieve all the data for a particular engineering or Flash calculation. Following this file search, any of the properties requested by the precalls for any of the components can be recovered any number of times. It is not necessary to keep the same component mixtures as the precall, — any subset of the components can be defined.

For those cases where some data is not available on the archive for either a component or a property, there is a facility provided for adding the missing data to the common area in CHEMCO by means of subroutine COMPAD. This means that further calculation can proceed as through all the data was available on the archive without the need to modify the archive itself.

Also included in the data bank system is a set of Flash routines which handle 6 different types of Flash calculations with options:

    (1)   with or without enthalpy balance
    (2)   ideal WILSON or UNIFAC VLE data.

The list of components available on the bank is given in Table 1, the physical properties are listed in Table 2 and the Flash calculations are described in Table 3.

Also provided with the data bank system is a main program for retrieval of data from the bank (GETDAT) and a program and data for generating the archive (GEN).

GETDAT enables the bank to be used simply to provide data for off-line use. It is capable of all the facilities offered to the on-line user, i.e. provision of parameters, point values, mixture properties and Flash calculations. The program is designed for both batch and interactive use.

The hierarchical structure of the system and the procedure for generating the archive are described in the appendices.

3.   DATA RETRIEVAL FROM THE BANK

The program GETDAT retrieves data from CHEMCO and FLASH and presents the results in a convenient printed form for the user. The program is designed both for interactive use and use in batch mode. All methods shown in Tables 2 and 3, with all components in Table 1 can be retrieved.

Figures 1 and 2, together with the notes following these figures describe the input requirement for the program. Following these there are two examples of its use in batch mode.

Missing data can be added by means of COMPAD to enable mixture properties to be calculated when some components or data are not on the archive. Section 5 describes the input requirements for COMPAD.

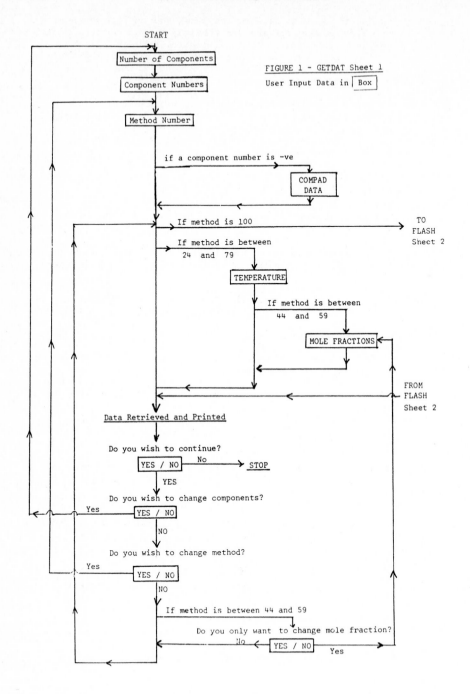

START

Number of Components

Component Numbers

FIGURE 1 - GETDAT Sheet 1
User Input Data in | Box |

Method Number

if a component number is -ve

COMPAD
DATA

If method is 100   TO
                                             FLASH
                                             Sheet 2

If method is between
24   and   79

TEMPERATURE

If method is between
44   and   59

MOLE FRACTIONS

                                             FROM
                                             FLASH
                                             Sheet 2

Data Retrieved and Printed

Do you wish to continue?

YES / NO   No   STOP

YES

Do you wish to change components?

Yes   YES / NO

NO

Do you wish to change method?

Yes   YES / NO

NO

If method is between 44 and 59

Do you only want to change mole fraction?

No   YES / NO   Yes

FROM   GETDAT   Sheet 1

FIGURE 2: GETDAT - Sheet 2

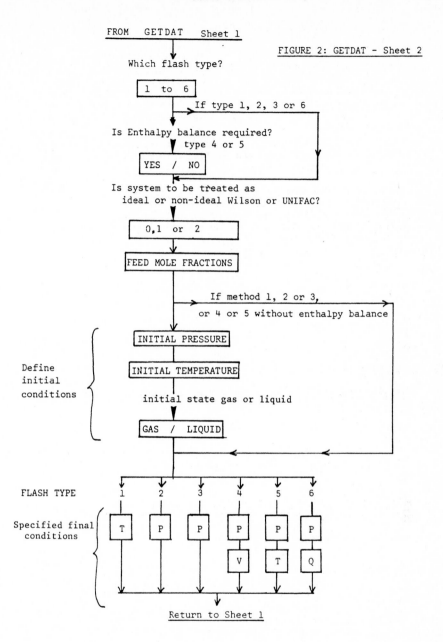

Which flash type?

1 to 6

If type 1, 2, 3 or 6

Is Enthalpy balance required?
      type 4 or 5

YES / NO

Is system to be treated as
ideal or non-ideal Wilson or UNIFAC?

0,1 or 2

FEED MOLE FRACTIONS

If method 1, 2 or 3,
or 4 or 5 without enthalpy balance

Define
initial
conditions

INITIAL PRESSURE

INITIAL TEMPERATURE

initial state gas or liquid

GAS / LIQUID

FLASH TYPE   1   2   3   4   5   6

Specified final
conditions   T   P   P   P   P   P

                                 V   T   Q

Return to Sheet 1

Notes on the Use of GETDAT - Figure 1 and 2

(1) All inputs to be given by the user are within the
     boxes on Figures 1 and 2.

(2) All data is free-format. Space or comma separates items. Names
     must be allowed 12 characters before numerical data follows.
     Where defaults are allowed, a simple carriage return (CR) suffices.

(3) Each input line is prompted for interactive use. In batch
     mode the input logic chart must be followed carefully to
     determine the data input sequence.

(4) New components or missing data are entered with COMPAD
     (see Section 5 for input requirements for COMPAD). As
     with normal COMPAD use, two following lines, each with
     -1 return command to GETDAT. COMPAD is called by setting the
     last component number negative.

(5) GETDAT operates with all methods shown in Table 2,
     including the Flash calculations.

(6) Mixtures are limited to a maximum of 10 components.

GETDAT Examples (BATCH USE)

1. Provide the basic data for acetaldehyde (1) and acetone (3).

        2
        1 3
        1

    NØ

2. Determine the condition after adiabatic expansion (FLASH
    Type 6) to a final pressure of 1 bar of a mixture of water
    (52) / methanol (31) (0.8 mf H_2O). Initial conditions being
    liquid at 10 bar and 393 $^{\circ}K$ (using Wilson VLE).

        2
        52 31
        100

        6
        1
        0.8   0.2
        10.0
        393.0
        LIQUID
        1.0
        0.0
        NØ

EXAMPLE 3   (BATCH USE)

For ethanol/water, find the Wilson coefficients, Unifac groups and parameters, and compare mixture vapour pressures at 100°C, calculated by ideal, Wilson and Unifac VLE models.

The control cards for CDC usage are also given.

ROSE,4507,CM100000,CT9.

```
ATTACH, GETDAT.     ⎫
ATTACH, CHEMCO.     ⎪     Example of Control Cards
ATTACH, LIB.        ⎬     for CDC and CYBER Batch Use.
ATTACH, DBIN.       ⎪
LOAD,GETDAT,CHEMCO,LIB. ⎭
EXECUTE.
est.
    2
  22 52
  80

YES
NO
YES
  19

YES
NO
YES
  79

350.
YES
NO
YES
 100

    1
    0
  .5           .5
 373.
YES
NO
NO
    1
    1
  .5           .5
 373.
YES
NO
NO
    1
    2
  .5           .5
 373.
NO
```

OUTPUT TO EXAMPLE 1

EURECHA PHYSICAL PROPERTIES DATA BANK

HOW MANY COMPONENTS
WHAT COMPONENTS ARE IN YOUR MIXTURE
WHAT PROPERTIES DO YOU REQUIRE

METHOD NO. 1
          BASIC DATA

 COMPONENT   MOL-WT   MT.PT. B.PT.   HF(V)   LAT.HEAT FREE ENER HF(L)

 1 ACETALDEHYDE   44.1 149.8 293.6 -166475.   26876. -133398. -193325.
 3 ACETONE   58.1 177.7 329.7 -216845.   30257. -152616. -248905.
DO YOU WISH TO CONTINUE   (YES/NO)
N

        OUTPUT TO EXAMPLE 2

        EURECHA PHYSICAL PROPERTIES DATA BANK

        HOW MANY COMPONENTS
        WHAT COMPONENTS ARE IN YOUR MIXTURE
        WHAT PROPERTIES DO YOU REQUIRE

        WHICH FLASH TYPE --(1 TO 6)
        VLE CODE:/=IDEAL, 1=WILSON, 2=UNIFAC
        1
         WHAT IS THE COMPOSITION OF YOUR MIXTURE (MF)
        WHAT IS THE INITIAL PRESSURE OF YOUR SYSTEM (BAR)
        WHAT IS THE INITIAL TEMPERATURE OF YOUR MIXTURE (K)
        WHAT IS THE INITIAL STATE OF YOUR MIXTURE(GAS/LIQUID)
        L
        WHAT IS THE FINAL PRESSURE OF YOUR SYSTEM

        INPUT SUMMARY FLASH IDEAL ENTH
                         6   0   1
        WHAT HEAT INPUT ARE YOU GIVING TO THE FLASH (KW/MOL)
                        FLASH CALCULATION TYPE   6

                          INITIAL   FINAL
                                         LIQUID   VAPOUR

        52WATER   .8000   .8291   .4733
        31METHANOL   .2000   .1709   .5273

         TEMP. (K)   393.00   357.56
         PRESS. (BAR)   10.000   1.000
         VAP.FRACTION   0.0000   .0845

                 ENTHALPY CHANGE/MOL FEED   0.00KJ/MOL

        DO YOU WISH TO CONTINUE   (YES/NO)
        N

Some Notes on Data Input

The subroutine library LIB contains a subroutine written in 1966 standard
FORTRAN (Subroutine FREAD) which enables all data to be read in effectively
in free-read.

The following points should be noted:

• Numerical data must be separated by a comma or blanks

• All data is read as real. Decimal points are only necessary
  when the data has a fractional part.

• Up to 40 data items can be packed on one 80 character line.
  Continuation lines are indicated by an asterisk (*) as the
  last item on the line to be continued.

• If alphanumeric data (names) are requested as part of the
  data, they must occupy the first 12 positions of the line.
  Any data within this field will be considered to be part of
  the name and not recognised as data (in this sense the input
  is not completely free-read).

• Carriage return (CR), returning zero characters on the line
  is recognised and used for input control or returning default
  values as indicated by the prompts.

• "Break" is $$$$, which in a limited number of instances (as
  indicated by prompts) enables one to return to an earlier
  point in the program and to re-enter erroneous data.

4. DIRECT ACCESS BY ENGINEERING PROGRAMS

    The bank can be used for direct access by any engineering program to
provide data for any method.

Methods 1 - 18	provide coefficients for use in the engineering program.
Methods 25-38	provide pure component properties.
Methods 45-58	provide mixture properties for direct use.
Method 79	provides UNIQUAC parameters
Method 80	provides Wilson coefficients
Method 100	provides a series of types of flash calculations.

The engineering program calling subroutine must be provided with

(a) the following COMMON block:

     COMMON/BANK1/KOUT,IR,NASTRO,NTOT,NCOAC,KSTAM,CNAME(3,10)

(b) the following DIMENSION statement:

     DIMENSION PROP(10,10), NORD(10), X(10), Y(10), Z(10)

(c) the subroutine calls:

CHEMCO for methods 1 - 80 inclusive
or
FLASH for method 100.

Notes: (1) The control cards must load FLASH and CHEMCO subroutines and
             their attendant subroutines, in addition to the engineering
             program.

        (2) Unless the main program has tape number 6, 14 and 5 for the
             output, archive file and input respectively, the block data
             setting these must be overwritten by redefining the variables
             IW, NASTRO and IR in the engineering program (Batch use).

        (3) If the engineering program is designed for repeated execution
             with different components, then the common areas in CHEMCO which
             store the data should be initialised to prevent overflow by
             setting

                          NTOT = 0
                     and NCOAC = 0

             in the engineering program.

        (4) The component names are available in the engineering program
             in array CNAME(3.10), e.g. the Ith component is stored in
             CNAME(J,I) (J=1,3) in 3A4 format.

        (5) KSTAM is a flag for controlling error messages and is set to zero.

4.1   The CHEMCO Subroutine Call

Prior to actual data use in the engineering program, the
necessary coefficients must be listed in CHEMCO prior to extraction
by a series of "PRECALLS".

CALL CHEMCO (1,METHOD,NORD,NCP,X,Y,T,P,PROP,IER)

The only active items in the argument list during the pre-
call are:

1   the first item in the argument list
              denotes a 'pre-call'

NCP   number of components in the system.

NORD(I),   Component number of each component
I=1, NCP   in the system.

METHOD   the physical property method required
              (see Table 2).

This pre-call must be repeated for every method that is
to be required. The pre-call can cover all components with each
call, but only one method. It is most effective (though not
essential) to make all pre-calls from one point in the engineering
program, since this enables all the data to be loaded by one
search of the archive.

The components in the mixture can be modified with each pre-call to prevent trying to load properties not available on the bank which will not be used later. For example, a system may contain 5 components, the molecular weights of all 5 are required, but only 3 will be significant in a later distillation. The pre-call for method 1 will have NCP=5, and that for the method 80 has NCP=3.

The properties are supplied to the engineering program by the CHEMCO call:

CALL CHEMCO (2,METHOD,NORD,NCP,X,Y,T,P,PROP,IER)

All items are now active:

2   denotes a call and not a pre-call.

METHOD   denotes the physical property method
               required.

NCP   number of components in mixture.

NORD(I)   component numbers of the components
               in the mixture.

X   liquid mole fractions of mixture.

Y   vapour mole fractions of mixture.

T   temperature of mixture.

P   pressure of mixture.

PROP(I,J)   returned property or coefficients
               (see Table 2)

IER   error flag $\neq$ 0 when CHEMCO cannot
               complete the request.
               NB: On the first call following the pre-calls,
               the archive will be searched for all pre-
               calls. Hence only at this point do error
               messages appear when any data is not available
               on the bank.

4.2 The FLASH Subroutine Call

For the Flash calculations (method 100 of Table 2 and Table 3) it is necessary for the engineering program to call the flash sub-routine, and not CHEMCO directly - see system hierarchy, Appendix 2. In this case, the CHEMCO pre-call and call are replaced by the FLASH pre-call and call.

The FLASH pre-call is:

CALL FLASH (1,IFL,NORD,NCP,Z,TIN,PIN,VIN,X,Y,T,P,V,Q,IER,IDEAL)

The only active variables are:

1   which denotes a <u>pre-call</u>

IFL   the Flash type required (see Table 3).
            Note that -ve means no enthalpy balance and
                        +ve with enthalpy balance.

NCP   number of components in mixture

NORD(10)   the component numbers in the mixture

IDEAL   = 0 for ideality
            = 1 for non-ideal Wilson
            = 2 for non-ideal UNIFAC

    The Flash pre-call -like the CHEMCO pre-call- simply
prepares to read the archive. The archive is read at
the first call following the pre-call.

The Flash <u>call</u> is:

    CALL FLASH(2,IFL,NORD,NCP,Z,TIN,PIN,VIN,X,Y,T,P,V,Q,IER,IDEAL).

    It can be used to perform any required Flash calculation,
providing the complete data has been requested by the pre-call,
e.g. the pre-call must have requested a flash type requiring
enthalpy data and Wilson coefficients if these are at any
time are to be requested in a FLASH call.

The argument list is as follows:

2   denotes a FLASH <u>call</u>.

IFL   the FLASH type required (see Table 3).

IDEAL   0 = ideality assumed
            1 = Wilson coefficients for VLE data.
            2 = UNIFAC method for VLE data.

NCP   number of components in the mixture.

NORD   the component numbers in the mixture.

X(10) ⎫
Y(10) ⎪
T   ⎬   liquid and vapour compositions, temperature, pressure
P   ⎪   and fraction vapour at the flashed condition.
V   ⎭

Z(10) ⎫
TIN   ⎪   the inlet composition (mole fraction), temperature
PIN   ⎬   pressure and vapour fraction. Only 0.0 or 1.0 (i.e.
VIN   ⎭   all liquid or vapour) allowed for VIN.

Q   represents the enthalpy change involved and gives
            the energy to be supplied to the system in J/mol.

IER   error flag, error when IER ≠ 0.

---

*Note: A Flash <u>repeat call</u> is also available - with the first item in the argument list
being 3. This can be used following a <u>call</u>, providing the list of components in
the mixture remains unchanged. The repeats call is then much quicker than call
since there is no need to re-enter CHEMCO.

Which of the items in the argument list are input data,
and which are retrieved by the subroutine depend on the FLASH
type chosen. Table 3 summarizes the defined and calculated
variables for each FLASH type.

EXAMPLE OF ENGINEERING PROGRAM USE:

A program requires the molecular weights of Hexane (26) and Benzene (7)
and also the liquid molar volume of a 0.5 mf (molar fraction) mixture of
the two components at 300°K.

```
      COMMON/BANK1/KOUT,IR,NASTRO,NTOT,NCOAC,KSTAM,CNAME(3,10)
      DIMENSION PROP(10,10),NORD(10),X(10),Y(10),Z(10)

      NCP=2
      NORD(1)=26
      NORD(2)=7
C  PRECALLS FOR METHODS 1 AND 53
      CALL CHEMCO (1,1,NORD,NCP,X,Y,T,P,PROP,IER)
      CALL CHEMCO (1,53,NORD,NCP,X,Y,T,P,PROP,IER)

C  CALL TO RETRIEVE MWTS AND PUT IN AMW1 AND AMW2
      CALL CHEMCO (2,1,NORD,NCP,X,Y,T,P,PROP,IER)
      IF(IER.NE.0) STOP1
      AMW1 = PROP(4,1)
      AMW2 = PROP(4,2)
         .
         .
         .

C  CALL TO RETRIEVE LIQUID MOLAR VOLUME AND PUT IN V
      T=300.
      X(1)=0.5
      X(2)=0.5
      CALL CHEMCO(2,53,NORD,NCP,X,Y,T,P,PROP,IER)

      V=PROP (1,1)
         .
         .
         .
         .
```


EXAMPLE CALLING 'FLASH' WITHIN AN ENGINEERING PROGRAM

The adiabatic flashing of an 0.5 mf mixture of Methanol (31), Methylcyclo-
hexane (32) at 470°K and 20 bar down to 1 bar is to be performed. What is
the resulting liquid and vapour composition, temperature and the fraction
vapourised. Wilson VLE data to be used.

```
      COMMON/BANK1/KOUT,IR,NASTRO,NTOT,NCOAC,KSTAM,CNAME(3,10)
      DIMENSION PROP(10,10),NORD(10),X(10),Y(10),Z(10)
         .
         .
         .

      NCP  2
      NORD(1)=31
```

```
      NORD(2)=32
      IDEAL=1
      IFL=6
C     FLASH PRECALL
      CALL FLASH(1,IFL,NORD,NCP,Z,TIN,PIN,VIN,X,Y,T,P,V,Q,IER,IDEAL)

      Z(1)=0.5
      Z(2)=0.5
      TIN=470.
      PIN=20.
      VIN=0.
      P=1.0
      Q=0.0
      T=350.
C     FLASH CALL
      CALL FLASH(2,IFL,NORD,NCP,Z,TIN,PIN,VIN,X,Y,T,P,V,Q,IER,IDEAL)
      IF(IER.NE.0) STOP1

C     Required information can be printed out or used within the program.

      WRITE (KOUT,100) X(1), X(2), Y(1), Y(2), T, V
100   FORMAT (6F10.4)
           .
           .
           .
           .
           .
```

5.   TEMPORARY ADDITION OF MISSING COMPOUNDS OR DATA BY SUBROUTINE 'C O M P A D'

This subroutine temporarily adds new components, or fills in omissions of data of existing components to the data extracted from the bank, held in CHEMCO. It does not add data to the archive and so at the end of the computer run the data added by COMPAD is lost.

It therefore enables the bank to be temporarily extended to meed individual mixture requirements. Data on the archive cannot be replaced, even temporarily by COMPAD. To modify archive data a new component must be created.

Use with Engineering Programs

Further data can be added by simply inserting the statement

CALL COMPAD

either before or after the CHEMCO or FLAHS calls. COMPAD then expects data to be presented in the following manner:

	NO. OF ITEMS	DATA	NOTES
LINE 1	1	Component Number	- New component number (70 to 100 available). - Existing component number for filling in data omissions. - -1 signifies no more components, i.e. return to calling program.

NO. OF ITEMS	DATA	NOTES

LINE 2   8   - Component Name (12 characters)

- mol wt   Method 1 data: If any data is not to be
- M.pt   used, or if it has been loaded from the
- B.pt   archive, then the line can be left blank
- Heat of Formation
  (vapour)
- Latent heat of vapourisation
- Free energy of formation
- Heat of formation (liquid)

LINE 3   1   Method Number

- Method 100 not allowed, must be built up from methods 16 and, if necessary 80, 13, 17 and 18 or 19.

- Line 3 does not automatically introduce method 1. If method 1 is to be retrieved, it must be entered as a line 4.

- -1, 0 or <CR> denotes no further methods for this component. Go to line 1.

- Note that the enthalpy methods are not loaded by enthalpy constants, but by heat capacity constants and heats of formation (see Table 6).

Line 4   5 or 6   Parameters for Method
- Number of parameters
- The parameter values

Parameters as required by program GEN (see Appendix 1) If Method=1 this line should be omitted. If Method=80 replace line 4 by 4a, b and c.

Line 4a   1   Number of second components for which binary data will be presented.

Only for Method 80.

Line 4b   8   Pairs of numbers, second component number, and Wilson coefficient (I,J). Number of pairs given by line 4a.

Line 4c   8   Pairs of numbers, second component number and second Wilson coefficient (J,I).

After line 4 (or 4c), the program requires a further line 3. Hence all new data can be added by one call of COMPAD. Two lines:

       -1
       -1

return control to the calling program (see lines 3 and 1).

Use with GETDAT

   If there is a need to add data not on the archive, then this information
is transmitted to the program by setting the last component number negative.
This passes control to the COMPAD subroutine which requires data in exactly
the same way as for the engineering program use. When GETDAT is used interactively the prompts given by COMPAD greatly reduce the chance of error.

Example: Data to add NEW COMP, No. 75, to CHEMCO giving mol. wt., vapour
          pressure and the Wilson coefficient with component 54 $W_{(75,54)}$,
          and its binary $W_{(54,75)} =$

```
    75

    NEW          100.   273.0     373.0   0.0    0.0    0.0    0.0
    16
     5     10.0         3000.0  100.0  200.0   500.0
    80
     1
    54    4000.0
    54    3000.0
    -1
    -1
```

6. TABLES

TABLE 1: LIST OF COMPOUNDS ON THE BANK

Bank Reference Number / Name

    1 Acetaldehyde
    2 Acetic Acid
    3 Acetone
    4 Acetylene
    5 Acrolein
    6 Ammonia
    7 Benzene
    8 Butane
    9 Carbon-Dioxide
   10 Carbon-Monoxide
   11 Carbon Tetrachloride (CCl_4)
   12 Chlorine
   55 Chlorobenzene
   14 Chloroethane   (ethyl chloride)
   15 Chloroform
   17 Cyclohexane
   20 Ethane
   22 Ethanol
   21 Ethyl-Acetate
   23 Ethylene
   18 Ethylene-Dichloride (1,2-Cl-Ethane)
   24 Ethylene-Oxide
   13 Freon 22 (C-H-Cl-F_2)
   25 Heptane
   26 Hexane
   27 Hydrogen

Bank Reference Number / Name

28 Hydrogen-Chloride
29 Hydrogen-Sulphide
45 Isopropanol
34 Ketone (Methyl- Ethyl)(CH_3-C_2H_5-)
60 Maleic Anhydride
30 Methane
31 Methanol
16 Methyl chloride (CH_3Cl)
32 Methyl-Cyclohexane
19 Methylene Chloride (CH_2-Cl_2)
33 Methyl Ether
53 M-Xylene
58 Naphthalene
35 Naphthalene (1-CH_3-)
36 Naphthalene (2-CH_3-)
61 Naphthaquinone
56 Nitrobenzene
38 Nitrogen
39 Octane
40 Oxygen
57 p-Dichlorbenzene
41 Pentane
42 Phenol
59 Phthalic Anhydride
43 Propane
37 Propane (2-CH_3-)
44 Propene
46 Pyridine
54 P-Xylene
47 Sulphur Dioxide (SO_2)
63 Sulphuric Acid   (H_2SO_4)
62 Sulphur Trioxide (SO_3)
48 Tetrachlorethylene (CCl_2=CCl_2)
49 Toluene
50 Trichlorethylene (Cl_3-Ethylene)
51 Vinyl Chloride
52 Water

Notes:

1. See Table 4 for the list of missing pure component data. Unless specifically
    mentioned in Table 4, pure component data for the above components is available
    on the bank.

2. See Table 5 for the list of included binary Wilson coefficient data.

TABLE 2 - Physical Property Selection

Property	"Method" or Property Code	Results returned in Array PROP(J,I) (Engineering Program Use only)		Units
		Array Location	Information returned for component I, where I=1, NCP	
Basic Data	1	PROP(1,I),PROP(2,I),PROP(3,I)	Component Name (3A4 Format)	---
		PROP(4,I)	Molecular Weight	---
		PROP(5,I)	Melting Point	K
		PROP(6,I)	Boiling Point	K
		PROP(7,I)	Heat of Formation (vapour)	J/mol
		PROP(8,I)	Latent Heat of Vapourisation	J/mol
		PROP(9,I)	Free Energy of Formation	J/mol
		PROP(10,I)	Heat of Formation (liquid)	J/mol
Vapour Heat Capacity				
- constants alone	5	PROP(1,I), PROP(2,I) PROP(3,I), PROP(4,I)	A_1 and A_2 of Equation (C) Lower and Upper Temperature Limits to the Correlation (°K)	
- point values for pure components	25	PROP(1,I)	Point Value at requested T and P	
- point value for a mixture	45	PROP(1,I)	Point Value at requested T and P and Concentration	
Liquid Heat Capacity				
- constants alone	10	PROP(1,I), PROP(2,I) PROP(3,I) PROP(4,I)	A_1 and A_2 of Equation C Lower and Upper Temperature Limits to the Correlation (°K)	J/mol.K
- point values for pure components	30	PROP(1,I)	Point Value at requested T and P	
- point value for mixture	50	PROP(1,I)	Point Value at requested T, P and Composition	

Property	"Method" or Property Code	Results returned in Array PROP(J,I) (Engineering Program Use only)		Units
		Array Location	Information returned for Component I, where I=1, NCR	
Vapour Enthalpy				
- constants alone	17	PROP(1,I),PROP(2,I),PROP(3,I) PROP(4,I), PROP(5,I)	A_1, A_2 and A_3 of Equation A Lower and Upper Temperature Limits to the Correlation (°K)	J/mol
- point values for pure components	37	PROP(-,I)	Point Value at requested T and P	
- point value for mixture	57	PROP(1,1)	Point Value at requested T, P and composition	
Liquid Enthalpy				
- constants alone	18	PROP(1,I),PROP(2,I),PROP(3,I) PROP(4,I),PROP(5,I)	A_1, A_2 and A_3 of Equation A Lower and Upper Temperature Limits to the Correlation (°K)	J/mol
- point values for pure components	38	PROP(1,I)	Point Value at requested T and P	
- point value for mixture	58	PROP(1,1)	Point Value at requested T and P and composition	
Vapour Thermal Conductivity				
- constants alone	7	PROP(1,I),PROP(2,I),PROP(3,I) PROP(4,I),PROP(5,I)	A_1, A_2 and A_3 of Equation A Lower and Upper Temperature Limits to the Correlation (°K)	W/mK
- point values for pure components	27	PROP(1,I)	Point value at requested T and P	
- point value for mixture	47	PROP(1,1)	Point Value at requested T, P and Composition	

TABLE 2 (cont'd.)

Property	"Method" or Property Code	Results returned in Array PROP(J,I) (Engineering Program Use only)		Units
		Array Location	Information returned for Component I, where I=1, NCP	
Liquid Thermal Conductivity				
- constants alone	12	PROP(1,I),PROP(2,I),PROP(3,I) PROP(4,I), PROP(5,I)	A_1, A_2 and A_3 of Equation A Lower and Upper Temperature Limits to the Correlation (°K)	W/mK
- point values for pure components	32	PROP(1,I)	Point Value at requested T and P.	
- point value for mixture	52	PROP(1,1)	Point Value at requested T, P and Composition	
Vapour Viscosity				Ns/m^3
- constants alone	6	PROP(1,I), PROP(2,I) PROP(3,I), PROP(4,I)	A_1 and A_2 of Equation C Lower and Upper Temperature Limits to the Correlation (°K)	
- point values for pure components	26	PROP(1,I)	Point Value at requested T and P.	
- point value for mixture	46	PROP(1,1)	Point Value at requested T, P and Composition.	
Liquid Viscosity				Ns/m^3
- constants alone	11	PROP(1,I),PROP(2,I),PROP(3,I) PROP(4,I), PROP(5,I)	A_1, A_2 and A_3 of Equation B Lower and Upper Temperature Limits to the Correlation (°K)	
- point values for pure components	31	PROP(1,I)	Point Value at requested T and P	
- point value for mixture	51	PROP(1,1)	Point Value at requested T and P and Composition.	

Property	"Method" or Property Code	Results returned in Array PROP(J,I) (Engineering Program Use only)		Units
		Array Location	Information returned for Component I, where I=1, NCP	
Liquid Molar Volume				l/mol
- constants alone	13	PROP(1,I),PROP(2,I),PROP(3,I) PROP(4,I), PROP(5,I)	A_1,A_2 and A_3 of Equation A Lower and Upper Temperature Limits to the Correlation ($^\circ$K)	
- point values for pure components	33	PROP(1,I)	Point value at requested T and P	
- point value for mixture	53	PROP(1,1)	Point value at requested T and P and Composition.	
Vapour Pressure				bar
- constants alone	16	PROP(1,I),PROP(2,I),PROP(3,I) PROP(4,I), PROP(5,I)	A_1,A_2,A_3 of Equation B Lower and Upper Temperature Limits to the Correlation ($^\circ$K)	
- point values for pure components	36	PROP(1,I)	Point value at requested T and P	
- point value for mixture	56	PROP(1,1)	Point value at requested T, P and Composition (assuming ideality)	
Surface Tension				N/m
- constants alone	15	PROP(1,I),PROP(2,I),PROP(3,I) PROP(4,I), PROP(5,I)	A_1,A_2 and A_3 of Equation A Lower and Upper Temperature Limits to the Correlation ($^\circ$K)	
- point values for pure components	35	PROP(1,I)	Point value at requested T and P	
- point value for mixture	55	PROP(1,1)	Point Value at requested T, P and Composition.	

P R O P E R T Y	"Method" or Property Code	Information returned
UNIFAC Group data for components (group number and number of such groups)	PROP(J,I) J=1,10 I=component	19 — Group numbers and numbers of such groups in each component (stored sequentially)
UNIQUAC UNIQUAC parameters as predicted from the UNIFAC group contribution method	PARAM(I,J) QS(I) RS(I) These variables are transferred by common/MOL/.	79 — Component interaction parameters mol. surface area mol. volume parameters for each component based on group contributions

APPENDIX 3 details the UNIFAC groups and the available UNIFAC group interaction parameters that are loaded on the CHEMCO bank.

TABLE 2 (cont'd.)

Property	Method	Results returned in Array PROP (J,I) (Engineering Program Use only)		Units
		Array Location	Information returned	
Wilson Coefficients Matrix of Wilson Coefficients (J,1) I=1, NCP J=1, NCP for a mixture containing NCP components	80	PROP(J,I)	The Wilson Coefficients $\Lambda(J,I)$ corresponding to Equation D. I=1, NCP J=1, NCP	
Flash Calculations 6 Types of Flash with enthalpy balance and ideality / non-ideality options	100	Engineering Programs must CALL FLASH(..................) NOT CHEMCO - see instructions on Flash. The returned Flash argument list contains the following:		
		X	Final Liquid Compositions	mf
		Y	Final Vapour Compositions	mf
		T	Final Temperature	°K
		P	Final Pressure	bar
		V	Final Vapour Fraction (0 ÷ 1.0)	---
		Q	Enthalpy change - heat input to system.	J/mol

Notes to TABLE 2

1. See Tables 4 and 5 for restrictions due to
    missing data in the archive.

2. The interactive program GETDAT accesses all
    the methods in this table.

3. For on-line engineering program use, methods 1 - 80
    call subroutine CHEMCO, and method 100 requires to
    call subroutine FLASH. See Section 4 "Direct Access
    by Engineering Programs" for details.

4. The equations used to correlate the data are:

    __Equation A__ property $= A_1 + A_2T + A_3T$

    __Equation B__ property $= \exp (A_1 - \dfrac{A_2}{A_3+T})$

    __Equation C__ property $= A_1 + A_2T$

    __Equation D__ - The Wilson Equation

    $$\ln(\gamma_i) = 1 - \ln \left(\sum_{j=1}^{N} x_i \Lambda_{i,j} \right) - \sum_{k=1}^{N} \left(\frac{x_k \Lambda_{k,i}}{\sum_{j=1}^{N} x_j \Lambda_{k,j}} \right)$$

    where

    γ_i is activity coefficient of i.

    x_i is liquid mole fraction of i.

    and

    $\Lambda_{i,j} = \dfrac{V_j}{V_i} \exp(-W_{i,j}/RT)$ when $i \neq j$

    $W_{i,j}$ are the Wilson coefficients (J/mol)

    V_i is liquid molar volume of the component i

    R   is the gas constant (8.3144 J/mol K).

5. The maximum number of components in a mixture
    is 10.

TABLE 3   THE 6 TYPES OF FLASH CALCULATION

NOTE: The FLASH Calculation is Considered to be Method 100

TYPE (IFLASH CODE)	Description	Defined Conditions	Calculated Conditions
1	Vapour pressure and composition over a liquid	x, T	P, y
2	Boiling point of a mixture	P, V, x (V=0.0)	T^*, y, Q
3	Dew point of a mixture	P, V, y (V=1.00)	T^*, x, Q
4	Partial vapourisation	P, V, z	T^*, x, y, Q
5	Isothermal flash	P, T, z	V^*, x, y, Q
6	Generalised adiabatic flash	P, Q, z (Q=0 for adiabatic flash)	T^*, V, x, y

*Although an initial estimate must be defined to start the calculation.

Nomenclature

x   mole fraction in liquid phase

y   mole fraction in vapour phase

z   mole fraction in feed

T   temperature °K

P   pressure bar

V   fraction vapour

Q   enthalpy change, +ve for heat to be supplied

NOTES TO TABLE 3

1. Enthalpy Balances

These are optional for types 4 or 5, and essential for type 6.

If the option with enthalpy balances is chosen, then the pre-flash conditions must be defined so that the enthalpy change can be calculated.

Hence: P_{in} - initial pressure

T_{in} - initial temperature

V_{in} - initial vapour fraction (either 0.0 or 1.0)

must all be defined.

The enthalpy option is indicated in the argument
list for subroutine FLASH by the sign of IFLASH.

e.g. IFLASH = -2   -Type 2 without enthalpy balance.
      IFLASH = 4   -Type 4 with enthalpy balance.

2. Ideal / Nonideal Option

For teaching and demonstration purposes the ideal
solution treatment is recommended. Computer time is
less and the range of components is not limited.

Nonideal treatment demonstrates azeotropic behaviour
and gives "correct" results.

This option is indicated in the argument list for
subroutine FLASH by:

IDEAL = 0   -ideal treatment
IDEAL = 1   -nonideal treatment Wilson
IDEAL = 2   -nonideal treatment UNIFAC

TABLE 4: MISSING PURE COMPONENT DATA

The archive does not contain the following data:

Property	Missing for Components No.:	Methods affected
Liquid heat capacity Liquid enthalpy Heat of formation of liquid	complete	10, 30, 50 18, 38, 58 100 (with enthalpy balance)
Vapour thermal conductivity	1-7, 9, 10, 12, 13, 17, 21, 24-29, 32-42, 45-49, 53, 54, 57,63	7, 27, 47
Liquid thermal conductivity	1-7, 9, 10, 12, 13, 16, 17, 21, 24-29, 32-43, 45-49, 51, 53, 54,57,62	12, 32, 52
Surface tension	34, 35, 36, 48,58-61	15, 35, 55
Liquid and Vapour viscosities	57-61, 63 (L)	6, 26, 46, 11 31, 51
UNIFAC component definitions by groups	4, 6, 9, 10, 12, 13, 27-29, 38, 40, 47, 48, 50, 51, 59-63 Additionally, not all group interactions are known, see appendix 3	19, 79 100 (ideal=2)

TABLE 5: WILSON BINARY COEFFICIENTS ARE AVAILABLE
          FOR THE FOLLOWING BINARY PAIRS

7	8	22	52
7	18	22	53
7	22	22	54
7	25	25	31
7	31	25	32
7	32	25	35
7	35	25	36
7	36	25	49
7	37	25	53
7	49	25	54
7	53	31	32
7	54	31	35
8	18	31	36
8	22	31	37
8	31	31	49
8	32	31	52
8	35	31	53
8	36	31	54
8	37	32	35
8	49	32	36
8	53	32	37
8	54	32	49
11	15	32	53
18	22	32	54
18	25	35	37
18	31	35	49
18	32	35	53
18	35	35	54
18	36	36	37
18	37	36	49
18	49	36	53
18	53	36	54
18	54	37	49
22	25	37	53
22	31	37	54
22	32	49	53
22	35	49	54
22	36		
22	37		
22	49		

Hence, method 80 and method 100 - Wilson   are limited
to the above mixtures, unless COMPAD is used to add missing data.

Method 100 using the ideal option can be used for any mixture.

CHEMCO Manual
APPENDIX 1:   GENERATION OF THE ARCHIVE

The archive for the data bank is built up from the
supplied parameters by the program GEN.

The data must be supplied in the following way:

(1) The number of components in the bank (I3)

(2) A card for every component, containing the
     component number and name in I3, 3A4 format.

(3) A -1 card to denote the end of the names (I3).

(4) The data cards, with the format 2I3, 5E13.5,
     3X, 1A1.

     For pure component properties, the input is:

     -compound number ⎫
     -method number   ⎬ defined by Table 6
     -coefficients   ⎭

     For binary data (Wilson coefficients) the
     input is:

     -first compound number
     -second compound number
     -2 Wilson parameters (W_{12} and W_{21})
     -'V' in column 75 to denote binary data.

     The order in which these cards are presented is
     immaterial, except that later cards update earlier
     cards.

(5) -1 indicates the end of the data (I3).

The program GEN writes the archive on file DBIN (tape
unit 4) in binary form as required by CHEMCO. For executions
GEN requires two temporary scratch file SCR (tape units 8 and 9)
which are not required after execution.

The input is NOT free-format, the specified fields must be strictly
adhered to. This appendix has been written for card data input. The
input can equally well be from a data file.

Structure of the Archive

(a) 1st Record: Gives the number of components in the bank (NSPE)
                  and total number of records (NREC).

(b) Next NSPE records give: Component number (I)
                              Number of Wilson coefficients $W(I,J)$
                              available (NW).

     Next NW pairs are: second component number (J)
                         Wilson coefficients $W_{(I,J)}$.

TABLE 6: PHYSICAL PROPERTY DATA FORMATS FOR ARCHIVE GENERATION

Method Input GEN	Methods made available in CHEMCO	Property	Units	Equations	Coefficients read in from Cards
1	1	molecular weight	-	-	A_1
		melting point	K	-	A_2
		boiling point	K	-	A_3
2	1	heat of formation of vapor	J/mol	-	A_1
		latent heat of vaporisation	J/mol	-	A_2
		free energy of formation	J/mol	-	A_3
		heat of formation of liquid	J/mol	(	A_4
5	5,25,45,17,37,57	vapor heat capacity	J/mol.K	c	$A_1\ A_2\ T_L\ T_u$
6	6,26,46	vapor viscosity	N s/m^3	c	$A_1\ A_2\ T_L\ T_u$
7	7,27,47	vapor thermal conductivity	W/m.K	a	$A_1\ A_2\ A_3\ T_L\ T_u$
10	10,30,50,18,38,58	liquid heat capacity	J/mol.K	c	$A_1\ A_2\ T_L\ T_u$
11	11,31,51	liquid viscosity	Ns/m^2	b	$A_1\ A_2\ A_3\ T_L\ T_u$
12	12,32,52	liquid thermal conductivity	W/m.K	a	$A_1\ A_2\ A_3\ T_L\ T_u$
13	13,33,53	liquid molar volume	m^3/kmol	a	$A_1\ A_2\ A_3\ T_L\ T_u$
15	15,35,55	surface tension	N/m	a	$A_1\ A_2\ A_3\ T_L\ T_u$
16	16,36,56	vapor pressure	bar	b	$A_1\ A_2\ A_3\ T_L\ T_u$

Equations: (a) property $= A_1 + A_2t + A_3t^2$

(b) $= \exp(A_1 - A_2/(t+A_3))$ $T_L \leq t_i \leq T_u$

(c) $= A_1 + A_2t$

| 19 | 19, 79 | Unifac group Nos. and No. of each group to define the component. | K | | |

Pairs of integer numbers packed into the same format, e.g. acetone, comp. 3, for method 19 has 1 gp No. 1, and 1 gp No. 18.

3 1901011801 0000.000000000000.   (2I3,5E13.5)

2 pairs

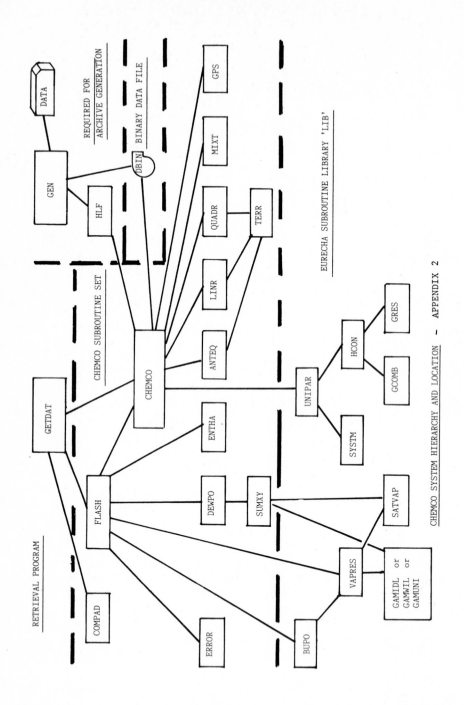

REQUIRED FOR
ARCHIVE GENERATION

DBIN BINARY DATA FILE

CHEMCO SUBROUTINE SET

EURECHA SUBROUTINE LIBRARY 'LIB'

RETRIEVAL PROGRAM

CHEMCO SYSTEM HIERARCHY AND LOCATION - APPENDIX 2

(c) <u>Next NSPE records give:</u> Name plus the 7 constants of method 1,
                            followed by 100 integers. Each integer is
     a pointer defining the location of the constants (-1000) for each
     method, e.g. the 7th pointer being 1013 means the 13th record
     after the component records gives the parameters for method 7 for
     the component in question.
     -1 denotes constants (or method) not available.
     0 denotes no constants required for that method.

(d) <u>Remaining records:</u> These are the constants for the methods and compo-
                            nents specified by the pointers in (c) above. Each
     record contains: - record number
                      - number of parameters (NP)
                      - followed by NP numbers, i.e.: the parameters of the
                        method.

The output of GEN gives a print-out of the archive formed.

<u>Enthalpy Data</u>

     To maintain consistency between thermal capacities of
liquid and vapour, latent heat and heat of formation of liquid
and vapour, and enthalpies, the following convention has been
used:

$$\text{Enthalpy} = \text{Heat of Formation} + \int_{25}^{t} C_p \, dt$$

To obtain strict consistency, liquid and vapour specific heat
the heat of formation of the vapour at 25° and the latent heat (L)
at the boiling point are taken as data and the heat of formation
of liquid is calculated from them as follows:

$$(H_f)_L = (H_f)_V + \int_{25}^{B_p t} (C_p)_V dt - L - \int_{25}^{B_p +} (C_p)_L \, dt$$

This calculation is performed by the subroutine HLF during the
generation of the data file.

APPENDIX 2:   CHEMCO SYSTEM HIERARCHY AND LOCATION - See figure opposite.

APPENDIX 3:   <u>DETAILS OF THE UNIFAC METHOD</u>

     The group definitions and available group parameters are given in the
following tables. These have been taken from the UNIDIST distillation program
manual by:

          Aage FREDENSLUND et al., Technical University of Denmark,
          Lyngby, October 1978.

     A complete description of the UNIFAC method is to be found in "The UNIFAC
Method", Fredenslund, Rasmussen and Gmehling, Elsevier, 1977.

     Note that the "Main Groups" are those from which the group interaction
parameters are determined (Table 2), and the sub-group number is that which
is used to define the component, and is the group required by method 19.

TABLE 7 : GROUP VOLUME AND SURFACE-AREA PARAMETERS.

Main Group	Sub Group	No	R_k	Q_k	Sample Group Assignments
1 "CH_2"	CH_3	1	0.9011	0.848	Butane: 2 CH_3, 2 CH_2
	CH_2	2	0.6744	0.540	
	CH	3	0.4469	0.228	2-Methylpropane: 3 CH_3, 1 CH
	C	4	0.2195	0.000	2,2-Dimethylpropane: 4 CH_3, 1 C
2 "$C=C$"	$CH_2=CH$	5	1.3454	1.176	1-Hexene: 1 CH_3, 3 CH_2, 1 $CH_2=CH$
	$CH=CH$	6	1.1167	0.867	2-Hexene: 2 CH_3, 2 CH_2, 1 $CH=CH$
	$CH=C$	7	0.8886	0.676	2-Methyl-2-butene: 3 CH_3, 1 $CH=C$
	$CH_2=C$	8	1.1173	0.988	2-Methyl-1-butene: 2 CH_3, 1 CH_2, 1 $CH_2=C$
3 "ACH" *	ACH	9	0.5313	0.400	Benzene: 6 ACH
	AC	10	0.3652	0.120	Styrene: 1 $CH_2=CH$, 5 ACH, 1 AC
4 "$ACCH_2$" *	$ACCH_3$	11	1.2663	0.968	Toluene: 5 ACH, 1 $ACCH_3$
	$ACCH_2$	12	1.0396	0.660	Ethylbenzene: 1 CH_3, 5 ACH, 1 $ACCH_2$
	$ACCH$	13	0.8121	0.348	Cumene: 2 CH_3, 5 ACH, 1 $ACCH$
5 "OH"	OH	14	1.0000	1.200	2-Butanol: 2 CH_3, 1 CH_2, 1 CH, 1 OH
6	CH_3OH	15	1.4311	1.432	Methanol: 1 CH_3OH
7	H_2O	16	0.92	1.40	Water: 1 H_2O

*Note A denotes benzyl ring

TABLE 7 continued:

Main	Group	No.			Example
8	$ACOH$	17	0.8952	0.68	Phenol: 5 ACH, 1 ACOH
9 "CH_2CO"	CH_3CO	18	1.6724	1.488	Ketone group is 2nd carbon; 2-Butanone: 1 CH_3, 1 CH_2, 1 CH_3CO
	CH_2CO	19	1.4457	1.180	Ketone group is any other carbon; 3-Pentanone: 2 CH_3, 1 CH_2, 1 CH_2CO
10	CHO	20	0.9980	0.948	Acetaldehyde: 1 CH_3, 1 CHO
11 "$COOC$"	CH_3COO	21	1.9031	1.728	Butyl acetate: 1 CH_3, 3 CH_2, 1 CH_3COO
	CH_2COO	22	1.6764	1.420	Butyl propanoate: 2 CH_3, 3 CH_2, 1 CH_2COO
12 "CH_2O"	CH_3O	23	1.1450	1.098	Dimethyl ether: 1 CH_3, 1 CH_3O
	CH_2O	24	0.9183	0.780	Diethyl ether: 2 CH_3, 1 CH_2, 1 CH_2O
	$CH-O$	25	0.6908	0.468	Diisopropyl ether: 4 CH_3, 1 CH, 1 CH-O
	FCH_2O	26	0.9183	1.1	Tetrahydrofuran: 3 CH_2, 1 FCH_2O
13 "CNH_2"	CH_3NH_2	27	1.5959	1.544	Methylamine: 1 CH_3NH_2
	CH_2NH_2	28	1.3692	1.236	Propylamine: 1 CH_3, 1 CH_2, 1 CH_2NH_2
	$CHNH_2$	29	1.1417	0.924	Isopropylamine: 2 CH_3, 1 $CHNH_2$
14 "CNH"	CH_3NH	30	1.4337	1.244	Dimethylamine: 1 CH_3, 1 CH_3NH
	CH_2NH	31	1.2070	0.936	Diethylamine: 2 CH_3, 1 CH_2, 1 CH_2NH
	$CHNH$	32	0.9795	0.624	Diisopropylamine: 4 CH_3, 1 CH, 1 CHNH
15	$ACNH_2$	33	1.0600	0.816	Aniline: 5 ACH, 1 $ACNH_2$

TABLE 7 continued:

Group	Subgroup	No.	R_k	Q_k	Example
16 "CCN"	CH_3CN	34	1.8701	1.724	Acetonitrile: 1 CH_3CN
	CH_2CN	35	1.6434	1.416	Propionitrile: 1 CH_2, 1 CH_2CN
17 "COOH"	COOH	36	1.3013	1.224	Acetic acid: 1 CH_3, 1 COOH
	HCOOH	37	1.5280	1.532	Formic acid: 1 HCOOH
18 "CCl"	CH_2Cl	38	1.4654	1.264	1-Chlorobutane: 1 CH_3, 2 CH_2, 1 CH_2Cl
	CHCl	39	1.2380	0.952	2-Chloropropane: 2 CH_3, 1 CHCl
	CCl	40	1.0060	0.724	2-Chloro-2-methylpropane: 3 CH_3, 1 CCl
19 "CCl_2"	CH_2Cl_2	41	2.2564	1.988	Dichloromethane: 1 CH_2Cl_2
	$CHCl_2$	42	2.0606	1.684	1,1-Dichloroethane: 1 CH_3, 1 $CHCl_2$
	CCl_2	43	1.8016	1.448	2,2-Dichloropropane: 2 CH_3, 1 CCl_2
20 "CCl_3"	$CHCl_3$	44	2.8700	2.410	Chloroform: 1 $CHCl_3$
	CCl_3	45	2.6401	2.184	1,1,1-Trichloroethane: 1 CH_3, 1 CCl_3
21	CCl_4	46	3.3900	2.910	Tetrachloromethane: 1 CCl_4
22	ACCl	47	1.1562	0.844	Chlorobenzene: 5 ACH, 1 ACCl
23 "CNO_2"	CH_3NO_2	48	2.0086	1.868	Nitromethane: 1 CH_3NO_2
	CH_2NO_2	49	1.7818	1.560	1-Nitropropane: 1 CH_3, 1 CH_2, 1 CH_2NO_2
	$CHNO_2$	50	1.5544	1.248	2-Nitropropane: 2 CH_3, 1 $CHNO_2$
24	$ACNO_2$	51	1.4199	1.104	Nitrobenzene: 5 ACH, 1 $ACNO_2$

The UNIFAC method does not have a complete set of sub-groups, and so some components cannot be described by subgroups and so cannot be handled by UNIFAC $-H_2$ for example.

Even if all components can be described by known sub-groups, the UNIFAC method can only be applied if all the resulting main group interaction coefficients (each main group with every other main group in the system) are known. Table 8 shows which ones are known to the bank, and so can be used to determine whether a particular system can be handled by UNIFAC. The parameter set in the CHEMCO bank is the 1978 updated set reported by Skjold-Jørgensen et al., Ind. Eng. Chem. Proc. Des. Dev., **18**, p. 714 (1979).

TABLE 8
THE UNIFAC Group Interaction Parameter Availability Table

('na' denotes parameters not available for that binary pair)

		1 CH2	2 C=C	3 ACH	4 ACCH2	5 OH	6 CH3OH	7 H2O	8 ACOH	9 CH2CO	10 CHO	11 COOC	12 CH2O
1	CH2												
2	C=C								na		na	na	
3	ACH										na		
4	ACCH2										na		
5	OH												
6	CH3OH								na				
7	H2O												
8	ACOH		na				na			na	na		na
9	CH2CO								na				
10	CHO		na	na	na				na			na	na
11	COOC		na								na		
12	CH2O								na		na		
13	CNH2			na					na	na	na	na	na
14	CNH								na	na	na		
15	ACNH2		na				na		na	na	na	na	na
16	CCN								na		na		na
17	COOH								na		na		
18	CCl								na			na	
19	CC12			na	na				na		na		
20	CC13								na		na		
21	CC14										na		
22	ACC1		na				na		na	na	na		na
23	CNO2								na		na	na	
24	ACNO2		na		na	na	na		na	na	na	na	na
25	CS2								na		na		
26	(C)3N				na		na		na	na	na	na	na
27	HCOOC		na	na	na			na	na	na	na		na
28	I	na		na		na		na	na		na		
29	Br	na					na	na	na		na	na	na
30	CH3SH	na		na				na	na		na	na	
31	CCOH					na					na		
32	Furfural	na					na	na	na		na		na
33	Pyridine	na								na	na	na	na
34	DOH	na					na	na		na	na		na

TABLE 8   contin.

		13 CNH2	14 CNH	15 ACNH2	16 CCN	17 COOH	18 CC1	19 CC12	20 CC13	21 CC14	22 ACC1	23 CNO2	24 ACNO2
1	CH2												
2	C=C			na							na		na
3	ACH							na					
4	ACCH2	na						na					na
5	OH												na
6	CH3OH					na					na		na
7	H2O												
8	ACOH	na	na	na	na	na	na	na	na		na	na	na
9	CH2CO	na	na	na							na		na
10	CHO	na	na	na	na	na		na	na	na	na	na	na
11	COOC	na		na			na					na	na
12	CH2O	na		na	na						na		na
13	CNH2			na	na	na	na	na	na	na		na	na
14	CNH			na	na	na	na	na	na			na	na
15	ACNH2	na	na		na	na	na	na	na		na	na	
16	CCN	na	na	na		na	na	na			na	na	na
17	COOH	na	na	na	na				na		na	na	na
18	CC1	na	na	na	na						na	na	na
19	CC12	na	na	na	na	na					na	na	na
20	CC13	na	na	na		na					na	na	na
21	CC14	na											
22	ACC1			na	na	na	na	na	na				na
23	CNO2	na	na	na	na	na	na	na	na				na
24	ACNO2	na	na		na	na	na	na	na		na	na	
25	CS2	na	na	na		na		na			na	na	na
26	(C)3N	na	na	na	na	na	na					na	na
27	HCOOC	na	na	na	na		na	na	na	na	na	na	na
28	I	na	na	na	na	na	na	na			na	na	
29	Br	na	na	na	na	na		na	na				na
30	CH3SH		na	na		na		na	na	na	na	na	na
31	CCOH					na						na	na
32	Furfural	na	na	na	na	na	na	na			na	na	na
33	Pyridine	na	na	na		na	na	na		na	na	na	na
34	DOH	na	na		na	na	na	na	na	na	na		na

TABLE 8 contin:

	25 CS2	26 (C)3N	27 HCOOC	28 I	29 Br	30 CH3SH	31 CCOH	32 Furfural	33 Pyridine	34 DOH
1 CH2										
2 C=C			na	na	na	na		na	na	na
3 ACH			na							
4 ACCH2		na	na	na		na				
5 OH							na			
6 CH3OH		na		na	na			na		na
7 H2O			na	na	na	na				na
8 ACOH	na	na	na	na	na	na		na		
9 CH2CO		na	na						na	na
10 CHO	na	na	na	na	na	na	na	na	na	na
11 COOC		na			na	na			na	
12 CH2O		na	na		na			na	na	na
13 CNH2	na	na	na	na	na			na	na	na
14 CNH	na	na	na	na	na	na		na	na	na
15 ACNH2	na	na	na	na	na	na		na	na	
16 CCN		na	na	na	na		na	na		na
17 COOH	na	na		na	na	na		na	na	na
18 CCl		na	na	na				na	na	na
19 CC12	na		na		na	na		na	na	na
20 CC13			na		na	na				na
21 CC14			na			na			na	na
22 ACCl	na	na	na	na		na		na	na	na
23 CNO2	na	na	na			na	na	na	na	
24 ACNO2	na	na	na	na	na	na	na	na	na	na
25 CS2	na	na	na	na	na	na		na	na	na
26 (C)3N	na		na	na	na	na	na	na	na	na
27 HCOOC	na	na		na	na		na	na	na	na
28 I	na	na	na		na	na	na	na	na	na
29 Br	na	na	na	na		na	na	na	na	na
30 CH3SH	na	na		na	na		na	na	na	na
31 CCOH	na	na	na	na	na	na		na	na	na
32 Furfural	na	na	na	na	na	na	na		na	na
33 Pyridine	na	na	na	na	na	na	na	na		na
34 DOH	na	na	na	na	na	na	na	na	na	

CHEMCO Manual
APPENDIX 4:   INSTRUCTIONS FOR LOADING CHEMCO DATA BANK

        CHEMCO   consists of two main programs, GEN and GETDAT, two subroutine
sets CHEMCO and LIB, and a data set to generate the data archive.

        Main program GEN has to be loaded with subroutine CHEMCO, and run with
the data set to produce the data archibe DBIN. This requires the assignment
of 5 files:

    - an input
    - an output
    - two scratch
    - and the final DBIN-file.

The GEN listing defines these files in more detail.

        To recover data from the bank, the main program GETDAT must be loaded
together with CHEMCO and LIB, and the data file DBIN must be available to
the program.

        If the bank is to be used to supply data to an engineering program, then
CHEMCO and LIB have to be loaded with the engineering program, and DBIN made
available. In this case, GETDAT is not required.

        Transfer to different machines should cause no problem since the coding
is strictly 1966 ANSI FORTRAN. The only necessary changes should be the assignment
of the three files: input, output, DBIN, which requires machine-specific instruction. The distributed coding contains CDC and PDP/VAX compatible statements.
On loading the program it must also be decided if batch, interactive or interactive + printer mode are required (modes 0 to 2, respectively).

    MODE 0   is straightforward batch operation requiring input and output
             files as units 5 and 6.
    MODE 1   is interactive, requiring input and output as the same unit,
             unit 5.
    MODE 2   allows for interactive input and operating dialogue, but with the
             results being stored on a separate file for later disposal to a
             printer (or directly on the printer output file). This requires
             units 5 and 7 to be defined.

Before compiling LIB, the appropriate DATA statements in BLOCK DATA, corresponding
to the required MODE must be activated by removing C from column 1. The unwanted
MODE DATA statements must be rendered inactive by inserting C in column 1 of each.

A multiple purpose version can be loaded as MODE 1, since in this mode the first
data item enables a mode switch to be made. Hence a zero as the first data item
converts the interactive program to batch use. When used in this way for batch
runs it must be remembered that the first item of every run must be a zero.

E U R E C H A

The European Committee for the Use of Computers in
Chemical Engineering Education

D I S T I L S E T

M A N U A L

────────────────────

(Program Version October 1984)

Manual Author: L.M. Rose
                   Technisch-Chemisches Labor
                   E.T.H. Zentrum
                   CH-8092 Zürich
                   SWITZERLAND.

CONTENTS   Page

1. INTRODUCTION

    Here is a set of three distillation computer programs which have been
specially developed for teaching purposes. They are related together in the
teaching aspects in that they form a natural progression from the simple
elementary graphical treatments to the advanced non-ideal multicomponent
situation normal to industry. The programs are also related in structure
and input requirements so that only one set of conventions is used for all
three programs. Physical property data can be entered manually or called from
the CHEMCO data bank. Ideal, Wilson and UNIFAC VLE models are available
where appropriate.

The three programs are based on programs that have been used for many years
for teaching but which were thoroughly revised and made compatible with each
other to form this set. Particular attention was given to the input, since
this is where most student mistakes are made. The input can be batch free-read
or interactive, either with supplementary printer or with all results being
presented on the terminal. Input checks are made wherever possible.

PROGRAM BINARY – Binary Distillation

This program is a plate-to-plate calculation of a binary distillation. It is
equivalent to the MacCabe-Thiele graphical design method in its simplest form,
but it extends this method to consider different total pressures, different
VLE models, to consider pressure drop and plate efficiency. The results are
presented as concentration and temperature profiles which serve as a good
introduction to multicomponent methods.

The program is a design program in that it determines the number of plates
for a required separation. It is so arranged that in one run it produces a
series of results at decreasing reflux ratios so that the number of plates
vs. reflux plot can be made.

Simple runs can be made using constant, given relative volatilities or any
of the VLE models available for the advanced distillation programs can be
called.

PROGRAM FENGIL – Short-cut Multicomponent Distillation

This is a coding of the well-known Feske-Underwood Gilliland method for determining the number of plates required for a given separation of key components
assuming that the components are ideal. This is useful as an introduction to
multicomponent distillation as it gives product distributions for adjacent
keys and is necessary to provide a starting estimate for rigorous distillation
studies.

For VLE data the program only needs relative volatilities but these can be
called up from the data bank if required and calculated from Antoine constants
for a given temperature, assuming ideality.

PROGRAM DISTIL – Multicomponent Distillation

This program is based on the widely used UNIDIST program of Fredenslund et al.
which was developed to demonstrate the UNIFAC VLE prediction method. The program
is a simplified multicomponent distillation in that it assumes ideality in the
vapour phase and constant molal overflow. This makes it particularly suitable
for teaching because the multicomponent principles can be demonstrated without
the excessive computation and data input necessary to carry out fugacity predictions and enthalpy balances on each plate.

The modifications to UNIDIST have been to the output to emphasize the demonstration of distillation rather than UNIFAC and to enable other VLE models to
be used, notably Wilson. Additionally the program has been coupled to the CHEMCO
data bank so the student can study distillation without first having to find,
input, correct and check physical property parameters before he can start his
distillation study. As with all EURECHA programs, input data standardisation
and checks have been incorporated.

The program is a simulation program which determines the ouptut compositions for a given column working at a given reflux ratio which given feeds at given plates with given distillate and side-stream take-off rates. The results are presented as composition, temperature and pressure profiles. Liquid activity coefficients for each component on each plate are also printed out.

The convergence algorithm used is the Naphtali-Sandholm method which gives adequately short computer times for small problems and is fairly robust. For teaching programs the algorithm chosen is not very important because demonstration problems are small and systems with convergence problems need not be chosen for teaching purposes.

Since this program is a simulation program, whereas the other two are design programs it enables the student to see the differences, to understand when each is possible and not possible and to learn to produce a design by intelligent repeated use of a simulation program.

PROGRAM SIZE AND TRANSFERABILITY

The total set of three programs has to be supplied with a subroutine library making a total of four sets of coding to be loaded with a total of about 3000 FORTRAN statements. If the data bank is to be used then this requires a further two files to be loaded containing the CHEMCO coding and the archive. For small machines it is possible to rearrange the library structure and load only the appropriate subroutines for the problem in hand. Depending on the facilities required these can be reduced to a bare minimum enabling the first two programs to be used on a very small machine.

Strict ANSI 66 FORTRAN has been used throughout and the programs are suitable for all machines, including 16 bit.

2. PROGRAM BINARY

This program determines the number of plates to perform a required separation for a binary continuous column. The program represents a simple constant molal overflow binary plate-to-plate calculation for a continuous distillation. The VLE data can be represented either by relative volatilities or by ideal, WILSON, UNIFAC or UNIQUAC VLE models.

From a specified bottom composition and relfux ratio, a plate-by-plate calculation is performed until the feed composition is reached. This defines the feed plate, and the point where the rectification mass balance equations are used. The plate-by-plate analysis then continues until the requested distillate composition is reached.

Reflux vs. number of plates and minimum reflux can be obtained by repeating cases in the same run for a series of decreasing ratios. Maximum plate data (line 2) can control the calculations so that the program stops when the number of plates is very high. Extrapolation of reflux vs. 1/number of plates will then give the minimum reflux ratio.

EXAMPLE of a Batch Run

```
ROSE,4507,CM100000,CT5.
ATTACH,BINARY.
ATTACH,LIB.                          example of CDC and CYBER
ATTACH,DBIN.                         control cards
ATTACH,CHEMCO.
LOAD,BINARY,LIB,CHEMCO.
EXECUTE.
eor
BENZENE/N-HEPTANE            UNIFAC FOR VLE
2
BENZENE         12.717       5017.34   41.82
N-HEPTANE       11.998       4719.9    22.55
373.
BANK             7
BANK            25
1.0,   .001,   0.90
TEST
0.1,   0.9,   0.5,   10.0,   1.0
0
```

line	Number of Items	Description
1	1	Title
2	3	$\left\{\begin{array}{l}-1 \text{ for relative volatility calculation} \\ 0 \text{ for ideal mixture} \\ 1 \text{ for VLE by Wilson equation} \\ 2 \text{ for UNIFAC} \\ 3 \text{ for UNIQUAC}\end{array}\right.$ - Max. no. of plates before giving up - (200 if left blank) - end recovery flag: =1 stop when no. of plates exceeded =0 recovers when no. of plates exceeded
3 if relative volatility is to be used	1	- relative volatility of mixture (α) at mean column temperature $= \dfrac{\text{SVP of light component}}{\text{SVP of heavy component}}$
if a VLE model is to be used then:	(3)	Vapour Pressure data (1) WITHOUT CHEMCO
NCOMP lines with 4 items		- Component Identification (12 positions) - A, B, C of the Antoine Equation or
NCOMP lines with 2 items		WITH CHEMCO - enter 'BANK' followed by 8 (or more)blanks - CHEMCO component identification number. (A negative number calls up COMPAD, to add data to the bank (see CHEMCO manual))
		VLE model data if IDEAL; this item is omitted, if WILSON,UNIFAC or UNIQUAC, go to the appropriate section.

WILSON
4a
	NCOMP(NCOMP-1)/2 lines with 3 items	WITHOUT CHEMCO
		- data identification (12 spaces, redundant)
		- two Wilson parameters (kJ/kmol), the following order must be used:
		Binary ternary N-components
		12 21 12 21 12 21
		             13 31 13 31
		             23 32 to
		                         1N N1
		                        then to
		                  (N-1)N N(N-1)

NCOMP(NCOMP-1)/2 lines with 3 items		or WITH CHEMCO - BANK followed by at least 8 blanks - 2 CHEMCO identification nos., entered in the order given by the table above.
5a	NCOMP lines with 4 items	WITHOUT CHEMCO - data identification (12 spaces, redundant) - A, B, C, of the liquid Molar Volume quadratic temperature model (2).
	NCOMP lines with 2 items	or WITH CHEMCO - BANK followed by at least 8 blanks - CHEMCO component identification no.

Item	No. of data items	Description
UNIFAC 4b	1 item	- mean temp (K) at which the coefficients are required.
5b	NCOMP lines each with a number of pairs of items.	WITHOUT CHEMCO - pairs of numbers, group no. and number of such groups; until all the different groups in the molecule are registered. The codes for the different groups are given in the CHEMCO manual. or
	NCOMP lines with 2 items	WITH CHEMCO - BANK followed by at least 8 blanks - CHEMCO component identification no.
UNIQUAC 4c	NCOMP(NCOMP-1)/2 lines with 4 items	Interaction parameters: - first component sequence no. - second component sequence no - first/second interaction parameter value - second/first interaction parameter value (see WILSON instructions for a systematic sequence).
5c	NCOMP lines with 3 items	R and Q data - the component sequence no. - R for that component - Q for that component UNIQUAC data cannot be called from the CHEMCO bank.
6	3	- Condensor system pressure (bars) - Pressure drop per plate (bars) - Overall plate efficiency (fraction) (blank card or ⟨CR⟩ gives 1 bar, no PD, 1.00 efficiency).
7	1	- sub-title
8	5	- required m.f. of light component in boiler - required m.f. of light component in distillate - m.f. of light component in feed - operating reflux ratio - heat content of feed (q), (fraction of feed as liquid).

Following this data a code in column 1 defines further runs:-

0= stop
1= repeat run with same data
2= repeat run with new data from subtitle (last 2 lines must be entered)
3= repeat run with all new data   (all new data must be entered)

NOTES

(1) The ANTOINE EQUATION form is:

$$\ln(P(\text{bar})) = A - \frac{B}{C + T(K)}$$

(2) The liquid molar volume model is:

$$V = a + B*T + C*T*T \quad (T \text{ in } (K), \ V \text{ in } M3/kmol)$$

(3) Physical property data can be inputted as direct data, or called from the CHEMCO data bank, or via temporary addition to the data bank, or by any mixture of all three methods.

3. PROGRAM 'FENGIL'

    The Fenske-Gilliland short-cut distillation method for obtaining initial estimates of distillation requirements for multicomponent mixture.

The Underwood equation

$$\sum_{i=1}^{n} \frac{x_{Fi}}{\frac{\alpha_i - \theta}{\alpha_i}} = 1 - q$$

where q is heat in feed = 0 for gas at dew point
                       = 1 for liquid at bubble point

is used to determine θ (Z GILL) by iteration with the Secant method. This value of θ is then used to determine the minimum L/D ratio:

$$\left(\frac{L}{D}\right)_{min} = \sum_{i=1}^{n} \frac{x_{Di}}{\frac{\alpha_i - \theta}{\alpha_i}} - 1$$

The light and heavy keys must be adjacent.
The number of plates at total reflux (N_m) are given by the Fenske equation.

$$N_m = \frac{\log[(x_{LK}/x_{HK})_D \ (x_{HK}/x_{LK})_B]}{\log(\alpha_{LK}/\alpha_{HK})_{AV}}$$

The Gilliland empirical correlation is then used to interpolate to determine the number of plates at a reflux ratio other than the minimum or total using the correlation of Eduljee (Hydr. Proc. 120, Sept. 1975)

$$\frac{N-N_m}{N+1} = 0.75 - 0.75 \left(\frac{\frac{L}{D} - \left(\frac{L}{D}\right)_{min}}{\frac{L}{D} + 1}\right)^{0.5688}$$

where N is the required number of plates for the working ratio L/D, and
      N_m is the number of plates at total reflux given by Fenske.

FENGIL

Use

The following data is required:

Notes	Number of Items		Description
	1		- Title

	3		- Number of components (max. 15) (NCOMP)
			- Source of VLE data
			  = 1 if vapour pressure is given by Antoine constants
			  = 0 if relative volatility is given.
			- mean temperature (K)

if Antoine constants supplied	2 or 4	first 12 positions reserved for name	- Antoine constants A, B and C for each component (see Note 1) $\ln(P\ \text{bar}) = A - B/(C + T\ K)$
or if relative volatility supplied	NCOMP		- a relative volatility, or pure component vapour pressure for each component (will be normalised)
	1		- sub-title
	1		- light key Number corresponding to order in which vapour pressure data is submitted
	NCOMP		- mole fraction of each component in the feed. In same order as vapour pressure input.
	5		- fraction light key recovered in bottoms $\left(\begin{array}{c}\text{see}\\ \text{Note 2}\end{array}\right)$ - fraction heavy key recovered in tops - required operating reflux ratio, as a ratio of the minimum reflux ratio (i.e. usually in the 1.1 to 2.0 range). - heat content of feed: 1.0 corresponds to liquid feed at bubble point 0.0 to vapour feed at dew point (q (i.e. fraction liquid in feed) - column feed rate (molar)

Following this data a code defines further runs: =0 stop; =1 repeat with same data; =2 return to sub-title; =3 return to title.

Notes

1. Format for Antoine constants corresponds to:
   Identification, component number or parameters A, B, C.
   If identification is BANK (no blanks before B), then the Antoine
   constants are taken from the CHEMCO data bank and the component
   number corresponds to the CHEMCO component identification number.
   A negative component number for the last component denotes that some
   data is not on the bank and COMPAD is then called to recieve new data.
   (see CHEMCO manual)

2. a) This recovery definition is NOT composition, but ratios of amounts
      fed to amounts recovered.
   b) These recoveries must be physically feasible and are usually less
      than 0.5.

Distribution of components other than the keys are again given by the
Fenske equation:

$$\frac{x_{D_i}}{x_{D_{LK}}} = \frac{x_{B_i}}{x_{B_{LK}}} \left(\frac{\alpha_i}{\alpha_{LK}} \right)^{N_m}$$

Determination of the feed point position is done by the Kirkbride method.

EXAMPLE 1   —with relative volatilities   (Batch run)

```
ROSE,4507,CM100000,CT5.        ⎞
ATTACH,FENGIL.                 │
ATTACH,LIB.                    │   example of CDC and CYBER batch
ATTACH,CHEMCO.                 │   control cards.
ATTACH,DBIN.                   │
LOAD,FENGIL,LIB,CHEMCO.        │
EXECUTE.                       ⎠
eor
3 COMPONENTS,  RELATIVE VOLATILITY
3 ,  0,  373.
4.0, 3.0, 1.0
TEST
1
.3, .3, .4
0.01, 0.01, 2.0, 1.0, 100.
0
```

EXAMPLE 2   —with vapour pressure from Antoine Coefficients

```
COEF GIVEN
     3    1  373.
  BENZ                  12.7170   5017.34   41.81
  NHEPT                 11.9976   4719.91   22.55
  ET:                   12.8916   4592.48  -16.46
  TRY
     1
   0.3        0.3       C.4
   0.01       0.01      2.00       1.0        100.
   0
```

EXAMPLE 3   —with Antoine Coefficients from the data bank.

```
CALLING BANK
   3     1   373.
BANK               7
NHEPT             11.9976   4719.91   22.55
BANK             22
TRY
   1
0.3          0.3        0.4
   0.01      0.01       2.00      1.0       100.
   O
```

EXAMPLE 4 —using bank with 1 component temporarily added

```
WITH COMPAD
   3     1   373.
BANK               7
NHEPT             11.9976   4719.91   22.55
BANK             -70
  70
NEW
  16
   5        12.8916   4592.48   -16.46    090      600.
  -1
  -1
TRY
   1
0.3          0.3        0.4
   0.01      0.01       2.00      1.0       100.
   O
```

4. PROGRAM 'DISTIL'

   This program simulates the performance of a given continuous distillation column separating a multicomponent mixture at a stated reflux ratio.

The program is based on the UNIDIST program of Fredenslund et al., Lyngby (Denmark) (see the UNIFAC method, Fredenslund, Rasmussen and Gmehling, Elsevier, 1977).

The program is a rigorous multicomponent distillation calculation using the Naphthali-Sandholm convergence procedure. Constant molal overflow is assumed, and only the liquid phase is treated for non-idealities.

The available VLE models are: — Ideal
                                — Wilson
                                — Unifac
                                — Uniquac.

DISTIL

No. of Data Items	Description
1	any title (e.g. describing system)
1	number of components (max. 10)
1	= 0 for ideal mixtures = 1 for non-ideal mixtures to be treated with Wilson = 2 for non-ideal mixtures to be treated with Unifac = 3 for non-ideal mixtures to be treated with Uniquac

PHYSICAL PROPERTIES (3)

                           Vapour Pressure data (1)

	WITHOUT CHEMCO
NCOMP lines with 4 items	- Component Identification (12 positions) - A, B, C of the Antoine Equation
or	
	WITH CHEMCO
NCOMP lines with 2 items	- enter 'BANK' followed by 8 (or more)blanks - CHEMCO component identification number. (A negative number calls up COMPAD, to add data to the bank (see CHEMCO manual))

                           VLE model data

         if IDEAL; this item is omitted,
         if WILSON,UNIFAC or UNIQUAC, go to the appropriate section.

WILSON	$NCOMP(NCOMP-1)/2$ lines with 3 items	WITHOUT CHEMCO - data identification (12 spaces, redundant) - two Wilson parameters (kJ/kmol), the following order must be used:

                                           Binary   ternary   N-components
                                           12 21   12 21   12 21
                                                    13 31   13 31
                                                      23 32   to
                                                                 1N N1
                                                             then to
                                                      (N-1)N N(N-1)

                      or

$NCOMP(NCOMP-1)/2$ lines with 3 items	WITH CHEMCO - BANK followed by at least 8 blanks - 2 CHEMCO identification nos., entered in the order given by the table above.

	WITHOUT CHEMCO
NCOMP lines with 4 items	- data identification (12 spaces, redundant) - A, B, C, of the liquid Molar Volume quadratic temperature model (2).
or	
	WITH CHEMCO
NCOMP lines with 2 items	- BANK followed by at least 8 blanks - CHEMCO component identification no.

No. of data items	Description
UNIFAC	
1 item	- mean temp (K) at which the coefficients are required.
NCOMP lines each with a number of pairs of items.	WITHOUT CHEMCO - pairs of numbers, group no. and number of such groups; until all the different groups in the molecule are registered. The codes for the different groups are given in the CHEMCO manual.
	or
NCOMP lines with 2 items	WITH CHEMCO - BANK followed by at least 8 blanks - CHEMCO component identification no.
UNIQUAC	
NCOMP(NCOMP-1)/2 lines with 4 items	Interaction parameters: - first component sequence no. - second component sequence no - first/second interaction parameter value - second/first interaction parameter value (see WILSON instructions for a systematic sequence).
NCOMP lines with 3 items	R and Q data - the component sequence no. - R for that component - Q for that component
	UNIQUAC data cannot be called from the CHEMCO bank.
2	— maximum change of temperature between iterations (Default = 10) — maximum fractional change in component flows between iterations (Default = 0.5)

NOTES

(1) The ANTOINE EQUATION form is:

$$\ln(P(bar)) = A - \frac{B}{C + T(K)}$$

(2) The liquid molar volume model is:

$$V = a + B*T + C*T*T \quad (T \text{ in } (K), V \text{ in } M^3/kmol)$$

(3) Physical property data can be inputted as direct data, or called from the CHEMCO data bank, or via temporary addition to the data bank, or by any mixture of all three methods.

Use

Number of Items	Description
2	— The definition of the initial profile and convergence details. $= 1$ for azeotropic distillation $= 0$ for normal profile (Default) Note: If an azeotropic distillate is involved, put $= 1$ and the heavy component should be the first component in the component list. - The negative exponential of convergence criteria (+6 giving error of 10^{-6} is default)
1	Any title (e.g. distillation description)
4	- Number of stages in the distillation (plate efficiency fixed at 1.00) (maximum 60). - Number of feeds to column. - Number of liquid side stream intermediate take-offs. - Number of vapour side stream intermediate take-offs.
6	- Number of moles of distillate - Reflux ratio (R/D) - Column pressure (bar) - Initial guess for top temperature (K) - Initial guess for bottom temperature (K) - Pressure drop across column (bar)
1	Feed line pairs. One pair per column feed: - Feed stage number.
NCOMP + 1	- Liquid fraction of feed stream (heat content of feed, q) - molar feed flow of each component (e.g. kmol/h).
1	The following 4 lines of data should be omitted if there are no side streams. Liquid product recovered line pairs (intermediate side streams): - Stage number from which product is removed.
1	- Molar flow rate of removed product (e.g. kmol/h.)
1	Vapour product removal card pairs (intermediate side streams): - Stage number from which product is removed.
1	- Molar flow rate of removed product (e.g. kmol/h).

Following this data , code in column 1 defines further runs:

      =0 stop
      =1 repeat run with same data
      =2 repeat run with new data from sub-title
      =3 repeat run with all new data (all data must be re-entered)

Example   - with VLE data predicted by UNIFAC (batch run)

Simulate the performance of a 10-plate distillation column for the mixture chloroform/chlorobromomethane/carbon tetrachloride, operating at a reflux of 5:1, UNIFAC is used for the VLE data.

```
ROSE, 4507,CM100000,CT9.    ⎞
ATTACH;DISTIL.              ⎟  example of CDC and CYBER control
ATTACH,LIB.                ⎟  cards for a batch run.
ATTACH,CHEMCO.             ⎟
ATTACH,DBIN.               ⎟
LOAD,DISTIL,LIB,CHEMCO.    ⎟
EXECUTE.                   ⎠
eor
```

```
   CF,CBM,CTC    UNIFAC
        3
        2
CF
CBM                        12.493Z   4833.91   52.8
CTC                        12.4682   4857.91   48.8
                           12.2540   4866.59   47.9
        350.
     CF              44 1
     CBM             38 1  57 1
     CTC             46 1
        10.      0.5
         0      6
     DIST TEST
        10    1
       0.2         5.        1.        333.       353.      0.1
         5
        1.0     .8        .2        .002
         0
```

EXAMPLE 2:   VLE with UNIFAC
             Some Vapour/Pressure and UNIFAC Groups
             from Data Bank

```
   N-OCTANE-ETHYLCYCLOHEXANE-ETHYLBENZENE-PHENOL
        4
        2
BANK           39
BANK           22
ATBZ               9.39952   3279.468    -59.9
BANK           42
430.
BANK           39
ETOH            1  1  2  6  3  1
ATBZ            1  1  9  5 12  1
BANK           42
10.        0.5
 0,0
 CHEMCO TEST
    20   1  0  0
    70.       8.     1.    388.    433.   0.
    10
0.0       25.0    25.0  25.0    25.0
 0
```

EXAMPLE 3:   VLE from UNIFAC
             All Data from Data Bank

BENZENE HEPTANE TOLUENE TEST
       3
       2
BANK   7
BANK   25
BANK   49
373.
BANK   7
BANK   25
BANK   49
     0,0
     0,0
DIST TEST
      10   1
  0.2   5.   1.   333.   353.   0.1
       5
  0.0   .8   .2   .002
   0

5.   INSTRUCTIONS FOR LOADING THE PROGRAM 'DISTILSET'

    The set contains 3 main programs: – BINARY
                                        – FENGIL
                                        – DISTIL
and a set of library subroutines LIB. It is necessary to load LIB with each of
the main programs.

If the data bank is to be used, then a second set of EURECHA programs, the
CHEMCOSET, must be obtained, and the subroutine CHEMCO loaded with the LIB
and the appropriate distillation program. The data archive must also have
been generated (see CHEMCO Manual) and must be available during the run as
data file DBIN.

The resulting set is of substantial size (4000 statements) and this may be too
big for smaller computers. In this case memory requirement can be reduced by
loading only those subroutines required for the problem in hand. To dispense
with the data bank releases CHEMCO and DBIN. To dispense with UNIFAC or WILSON
VLE models render further subroutines in LIB redundant which can then be
removed. In the limit, BINARY and FENGIL (with relative volatilities only) are
very small programs.

Transfer to different machines should cause no problem since the coding is
strictly 1966 ANSI FORTRAN. The only necessary changes should be the assignment of the 3 files INPUT, OUTPUT and DBIN, which requires machine-specific
instructions. The distributed coding contains CDC and PDP/VAX file handling
statements. On loading the program it must also be decided if batch, interactive
or interactive + printer mode are required (modes 0 to 2, respectively).

MODE 0   is straightforward batch operation requiring input and output
         files as units 5 and 6.
MODE 1   is interactive, requiring input and output as the same unit,
         unit 5.

MODE 2   allows for interactive input and operating dialogue, but with
the results being stored on a separate file for later disposal
to a printer (or directly on the printer output file). This
requires units 5 and 7 to be defined.

Before compiling LIB, the appropriate DATA statements in BLOCK DATA, corresponding to the required MODE must be activated by removing C from column 1.
The unwanted MODE DATA statements must be rendered inactive by inserting C
in column 1 of each.

A multiple purpose version can be loaded as MODE 1, since in this mode the
first data item enables a mode switch to be made. Hence a zero as the first
data item converts the interactive program to batch use. When used in this
way for batch runs it must be remembered that the first item of every run
must be a zero.

6. SOME NOTES ON DATA INPUT

The subroutine library LIB contains a subroutine written in 1966 standard
FORTRAN (Subroutine FREAD) which enables all data to be read in effectively
in free-read.

The following points should be noted:

● Numerical data must be separated by a comma or blanks

● All data is read as real. Decimal points are only necessary when the
data has a fractional part.

● Up to 40 data items can be packed on one 80 character line. Continuation
lines are indicated by an asterisk (*) as the last item on the line to be
continued.

● If alphanumeric data (names) are requested as part of the data, they must
occupy the first 12 positions of the line. Any data within this field will
be considered to be part of the name and not recognised as data (in this
sense the input is not completely free-read).

● Carriage return (<CR>), returning zero characters on the line is recognised
and used for input control or returning default values as indicated by the
prompts.

● "Break" is $$$$, which in a limited number of instances (as indicated by
prompts) enables one to return to an earlier point in the program and to
re-enter erroneous data.

E U R E C H A

The European Committee for the Use of Computers in
Chemical Engineering Education

"BATCH" BATCH DISTILLATION DESIGN PROGRAM
M A N U A L

(Program Version October 1984)

Manual Author: L.M. Rose
Technisch-Chemisches Labor
E.T.H. Zentrum
CH-8092 Zürich
SWITZERLAND.

CONTENTS

1.   THE DESIGN PHILOSOPHY

The types of batch distillation design equipment question posed to the designer usually takes the form either:

- what is the batch time when some existing equipment is used for a new duty?
  or
- what size equipment is required to complete a given batch within a specified time.

Both questions involve the performance of the total system, and are therefore best answered by simulating the total system.

The BATCH program simulates a total batch distillation system (see Fig. 1). - the number of plates and flow characteristics of the column, the boiler volume and area, the condenser area and the control system necessary to maintain on-specification product and full loading of the equipment. The simulation proceeds through each product fraction and intermediate fraction to obtain a final simulated total time (and total heat consumption).

To use BATCH as a design tool it is necessary to propose a distillation plant (specifying dimensions and operating procedure) and simulate the proposal to see what the total cycle time is, and what the major bottlnecks are.

Repeated simulations with modified designs enable the designer to achieve a satisfactory design using an evolutionary design procedure.

2. THE SIMULATED SYSTEM

Fig. 1 defines those parts of a batch distillation system that are included in BATCH. The components of the system are as follows:

Column: The column is defined by its diameter, giving the operating F factor $V \sqrt{\rho_g}$ kg$^{1/2}$ s^{-1} m$^{-1/2}$) for the internals and the pressure drop per theoretical stage at design conditions. Table I gives some typical data.

The effect of liquid hold-up is accounted for by specifying the column hold-up defined as liquid volume percent at operating conditions.

Boiler: Three types of boiler are simulated:

1. vertical jacketed cylinder,
2. horizontal jacketed cylinder,
3. unjacketed cylinder, all areas supplied as internal submerged surface.

The three types differ in the way the heat transfer surface is exposed to the boiling liquid and how this area reduces as the liquid volume is reduced. An overall heat transfer coefficient for submerged area must be supplied.

Additional boiler area is always assumed to be submerged, whereas jacketed area is liquid covered to an extent depending on liquid volume present. The boiler can be heated by any heating medium with a variable (or a fixed) temperature.

Condenser: This is simulated using a given overall heat transfer coefficient and a coolant of given inlet and outlet temperature. The active cooling area can thus be calculated and compared with the area supplied. The effect of column head hold-up can be simulated by defining the volume of liquid involved as "condenser hold-up".

Product Specification Control Loop

This control loop changes the reflux ratio to maintain the product specification. For intermediate fraction the control is inactive and the fraction is withdrawn at a fixed reflux defined by the user.

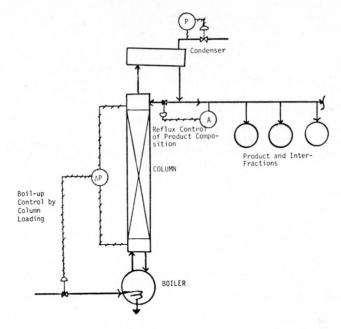

Fig. 1: The Simulated System

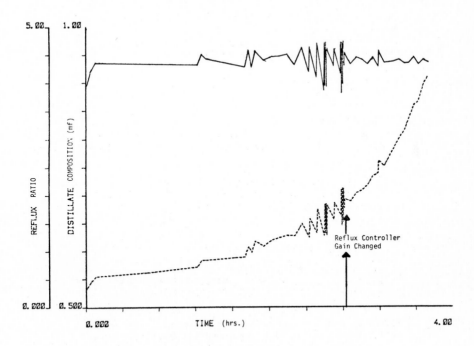

Fig. 2: Occurrence of Instability and its Removal

If the top product is designated as two-liquid phase, then
the light component is removed in preference to the other
components. (decantation). The reflux control then adjusts the
take-off rate so that it matches the quantity of light component
present in the overheads. This option is only available to
intermediate fractions.

Equipment loading control loop

Full loading of the column(or of the condenser, whichever is
limiting is maintained by controlling the heating medium
temperature to the boiler. When the heating medium temperature
reaches its maximum available temperature, then the control
loop automatically switches over to modify column head pressure
to maintain full loading.

3. INPUT DATA REQUIRED

The BATCH   program uses an input data file BATDAT which it prepares,
modifies, reads and writes prior to entering the calculation phase.

The data required fall into 4 categories: the physical properties of the
system, the description of the equipment, the definition of the operating
conditions and program control.

The input data is defined in Table 1 . In interactive mode the data input
is considerably simplified by the prompts which appear before every set of data
to be read in.

TABLE 1
INPUT DATA FOR PROGRAM BATCH

Item	No. of data items	Description
A) PHYSICAL PROPERTIES (3)		
1	1 line	- a title to define the physical system
2	2 items	- No of components (between 2 and 5) - give VLE model code: 0 ideal 1 Wilson 2 UNIFAC 3 UNIQUAC
		Vapour Pressure data (1)
3	NCOMP lines with 4 items	WITHOUT CHEMCO - Component Identification (12 positions) - A, B, C of the Antoine Equation or
	NCOMP lines with 2 items	WITH CHEMCO - enter 'BANK' followed by 8 (or more)blanks - CHEMCO component identification number. (A negative number calls up COMPAD, to add data to the bank (see CHEMCO manual))

Item	No. of data items	Description

VLE model data

if IDEAL; items 2 and 3 are omitted,
if WILSON,UNIFAC or UNIQUAC, go to the appropriate section.

WILSON
4a

NCOMP(NCOMP-1)/2
lines with 3 items

WITHOUT CHEMCO
- data identification (12 spaces, redundant)
- two Wilson parameters (kJ/kmol), the following
  order must be used:

Binary	ternary	N-components
12 21	12 21	12 21
	13 31	13 31
	23 32	to
		1N N1
		then to
		(N-1)N N(N-1)

or

NCOMP(NCOMP-1)/2
lines with 3 items

WITH CHEMCO
- BANK followed by at least 8 blanks
- 2 CHEMCO identification nos., entered in the
  order given by the table above.

5a

NCOMP lines
with 4 items

WITHOUT CHEMCO
- data identification (12 spaces, redundant)
- A, B, C, of the liquid Molar Volume quadratic
  temperature model (2).

or

NCOMP lines
with 2 items

WITH CHEMCO
- BANK followed by at least 8 blanks
- CHEMCO component identification no.

UNIFAC
4b

1 item

- mean temp (K) at which the coefficients are
  required.

5b

NCOMP lines each
with a number of
pairs of items.

WITHOUT CHEMCO
- pairs of numbers, group no. and number of
  such groups; until all the different groups
  in the molecule are registered. The codes for
  the different groups are given in the CHEMCO
  manual.

or

NCOMP lines with
2 items

WITH CHEMCO
- BANK followed by at least 8 blanks
- CHEMCO component identification no.

UNIQUAC

4c

NCOMP(NCOMP-1)/2
lines with 4 items

Interaction parameters:
- first component sequence no.
- second component sequence no
- first/second interaction parameter value
- second/first interaction parameter value
  (see WILSON instructions for a systematic
  sequence.)

Item	No. of data items	Description
5c	NCOMP lines with 3 items	R and Q data - the component sequence no. - R for that component - Q for that component

UNIQUAC data cannot be called of the CHEMCO bank.

6

Liquid enthalpy data (4)

WITHOUT CHEMCO

NCOMP lines of 4 data items	- data identification (12 spaces ,redundant) - A, B and C parameters of the enthalpy model quadratic in temperature (4).

or

WITH CHEMCO

NCOMP lines of 2 data items	- BANK followed by at least 8 blanks - CHEMCO component identification No.

7

Vapour enthalpy data

input exactly as item 4 but for vapour not liquid.

8

Liquid molar volumes

Input exactly as item 2b. Hence the data will have been entered twice it WILSON was chosen as the VLE model.

9

Molecular Weight

WITHOUT CHEMCO
- data identification (12 spaces, redundant)
- mol wt

NCOMP lines of 2 items   or

WITH CHEMCO
- BANK followed by at least 8 blanks
- CHEMCO component identification No.

B) DEFINITION OF EQUIPMENT

10	1 item	- title defining equipment

Item	No. of data items	Description

Boiler details:

| 11 | 1 line with 5 items | - boiler type code;
    1 - vertical jacketed cylinder
    2 - horizontal jacketed cylinder
    3 - unjacketed cylinder, all h.t. area
       to be supplied as additional area.
- boiler diameter (m)
- boiler length (height) (m)
- overall heat transfer coefficient for boiler ($kW/m^2 K$)
- additional heat transfer area to be supplied (m^2), as tubes, as opposed to the jacket. |

Condenser details

| 12 | 1 line with 3 items | - condenser area (m2)
- overall heat transfer coefficient for condenser (kW/m2 K)
- condenser and overheads hold-up (m^3) |

Column details

| 13 | 1 line with 6 items | - column diameter (m)
- column height (m)
- Number of theoretical stages
- F-Factor at design loading (6)
- operating P.D. per theoretical stage at design loading (mbar),(6)
- column hold-up (% column volume),(6) |

C) UTILITIES INFORMATION

| 14 | 1 line with 4 items | - Maximum heating medium temperature (K)
- Inlet cooling water temperature (K)
- Max. allowed outlet cooling water temp. (K)
- Maximum allowed total pressure (bar) |

D) OPERATING CONDITIONS

| 15 | 1 item | - title to define operation |

| 16 | 1 line with NCOMP + 2 | - Maximum distillation time (h) (7)
- Volume of initial charge (m^3)
  (if zero, full boiler is assumed)
- Mol fraction of each component (NCOMP items)
  (These are firstly normalised by the program) |

Item	No. of data items	Description

Product specification

| 17 | NCOMP lines of 2 items | - mol fraction lights spec. in product fraction (zero for first fraction)
 - mol fraction heavies spec. in product fraction (zero for heaviest fraction, i.e.boiler residue) |

Repeat this data for each product fraction.

Reflux operating policy (7)

| 18 | (NCOMP -1) lines of 4 items | - Starting reflux ratio for product fraction
 - finishing reflux ratio for product fraction
 - following inter-fraction reflux ratio.
 - Phase flag for inter-fraction:-
 0 = single liquid phase
 1 = two liquid phases (light product decanted) |

Repeated for all distillate product fractions.

NOTES

(1) The ANTOINE EQUATION form is:

$$\ln(P(bar)) = A - \frac{B}{C + T(K)}$$

(2) The liquid molar volume model is:

$$V = a + B*T + C*T*T \quad (T \text{ in } (K), V \text{ in } M^3/kmol)$$

(3) Physical property data can be inputted as direct data, or called from the CHEMCO data bank, or via temporary addition to the data bank, or by any mixture of all three methods.

(4) The Enthalpy model is:

$$H = A + B*T + C*T*T \quad (T \text{ in } (K), H \text{ in } kJ/kmol)$$

(5) The component order must correspond to the volatility sequence, the most volatile first.

(6) Table 2 gives some typical values.

(7) The program runs interactively enabling these data to be modified during the run.

(8) The program can be terminated at any point during the simulation by setting the integration step size to -1,when it is asked if any changes are needed.

(9) Some notes on data input:
The data input is effectively free read, and a <CR> returns default or previous read in data without change.

a) Numerical data separated by comma or blank.

b) All data assumed real. Decimal point only necessary to define a fractional part.

c) Up to 40 items per 80 character line can be read in. Continuation on to
   a new line requires a star(*) as the last item on the line to be continued.

d) If names are requested as part of the data, they must occupy the first
   12 positions of the line. Any data within this field will be considered
   to be part of the name and will not be recognised as data.

TABLE 2

TYPICAL DATA FOR SOME COLUMN INTERNALS

Internals	F-Factor at design operating conditions $kg^{1/2}_m{}^{-1/2}s^{-1}$	P.D. per tray at operating condition mbar	column Ht. for one theoretical tray m	Hold-up volume % of column %
25 mm Pall Rings	2.3	2.5	0.33	7
50 mm Pall Rings	2.5	2.4	0.46	7
Sieve tray	2.1	6.0	0.7	12
Sulzer structured packing	2.8	1.4	0.27	7

4. USE OF THE PROGRAM

On running the prepared data file with BATCH, prompts define further control
data which must be entered:

The first of these is the VLE source. For short-cut distillation calculations
a constant relative volatility (alpha) can be entered. Alternatively, the relative
volatility calculated from given VLE data for the boiler stage can be taken as
constant over the whole column. The third option is a full VLE calculation on each
plate.

The second item of control data is the quantity of information to be displayed.
Extended print-out gives a column concentration and temperature profile and some
additional heat load information at each integration step.

The third data item defines the number of integration steps to be completed
before stopping for a review of the conditions.

Having been given this information, the appropriate number of steps are simulated and tabulated on the screen (see example output). There is then the opportunity
to carry on for a further set of steps unchanged, or:

- to change to next fraction (i.e. override the strategy given in initial data)
- to change input from or to extended form,
- to alter the number of integration steps before stopping with questions,
- to change the gain in the temperature control/column loading control loop
  (should this control loop be unstable or sluggish),
- to change the operating reflux ratio. This is necessary if the reflux
  ratio is not controlling the specification, if it is wildly oscillating
  and manual correction is necessary.

- to change the gain in the reflux ratio / top specification loop, if
  this control is unstable or sluggish,
- to change the integration step length. This is defined as the fraction
  of light present in the boiler to be removed in one integration step.
  If -1 is entered here, the distillation stops and a final report is
  produced.
- to change the existing VLE method (given alpha, boiler alpha, or complete
  VLE).

This set of questions enables the engineer to control the course of the
distillation, in particular to overcome control instabilities and shortened
computer run-time. With well-chosen control gain parameters the total simulation
can run through using the pre-defined strategy, without changes at these question
intervals. Usually, however, at least the reflux gain must be modified at some
point to achieve a complete distillation.

After all distillation fractions have been simulated, a cumulative report
is produced summarising the distillation time, fraction composition and heat
requirement (see output example). Also given is the time to heat the boiler
contents to the boiling point, and the time at total reflux to establish
equilibrium before take-off commences (taken as a 5-fold boil-up of the quantity
of liquid held up in the column and condenser).

5. THE INTERACTIVE FACILITY

The interactive facility during the calculation is an essential feature of
the program for two reasons.

Firstly the calculation is unstable if incorrect values are used for the
controller gains. The difficulty of locating the correct parameter values is
overcome by the interactive facility which enables new values to be tried when
the existing values are proving to be unstable. Instabilities are also inherent
in the system when hold-up and pressure-drop are significant. Again, such instabilities can be removed by suitable choice of control parameters and step size.
Fig. 2 shows the occurrence of instability and its removal by gain parameter
adjustment.

Secondly, the design method requires the definition of a control strategy
before runs can be made. The interactive facility allows these strategies to be
modified in the light of immediate past results, so considerably reducing the
overall number of runs necessary.

Thirdly, considerable computer time can be saved by making use of the
short-cut VLE facilities, increasing the integration step length and terminating
abortive runs. Switching from short-cut to full method for VLE for checks during
the distillation indicating the magnitude of the error to be expected.

A particularly useful technique is to reduce the step length to be very small
to investigate the effect of parameters on various instabilities, or the importance
of the correct VLE. This holds the distillation at a virtual stand-still, but still
enables the effect to be fully investigated. Hence, the interactive facility
gives the design engineer a very good 'feel' for the problem he is solving.

6. AN EXAMPLE

4.0 m3 of a mixture of methanol/water/dimethyl formamide (0.5, 0.3, 0.2 mf, respectively) is to be batch distilled in a 400 mm diameter column packed with 3 m of 50 mm Pall rings (0.5 m) The boiler supplied is a vertical jacketed cylinder 2.75 m long and 1.4m diameter, with 2 m2 of internal coils. A water cooled condenser of 5 m2 surface is provided. The packing   hold-up is 7% column volume   , and the condenser system 0.03 m3. Steam is available for the boiler at 150°C, and cooling water 20°C should have a return temperature of 30°C.

Methanol product specification is 0.95 mf, water 0.95 mf, and the DMF in the boiler should contain less than 0.05 mf water. Heat transfer coefficients for both boiler and condenser are   2 kW/K m2. As a defined operating policy, product fraction starts with reflux ratios of 0.5 and finish when the reflux ratio 5.0. Intermediate fractions are removed with a reflux ratio of 4.0.

The following pages list the data input and give the results of the simulation.

```
MANUAL EXAMPLE    MEOH/H2O/DME
3  1
BANK                31
BANK                52
DMF               9.7478  3540.868  -62.77
12   21           -1234.9  3758.6
13   31           4348.4  -4541.8
23   32           3592.9  -763.1
BANK                31
BANK                52
DMF               0.0811    0.0   0.0
BANK                31
BANK                52
DMF                 0.0  0.0  0.0
BANK                31
BANK                52
DMF               30000.   0.0  0.0
BANK                31
BANK                52
DMF               0.0811    0.0   0.0
BANK                31
BANK                52
DMF               73.1
TEST
1  1.4  2.75  2.0  2.0
5.0  2.0  0.03
.40   3   6   2.5   2.4   7
423.   293.    303.    1.016
FIRST TRIAL
12.0,  4.0,   0.5,  0.3,  0.2,  293.
0.0,  0.05
0.025,  0.025
0.05, 0.0
0.5,  5.,  4., 0
0.5,  5.,  4., 0
```

MANUAL EXAMPLE   MEOH/H2O/DME

COMPONENTS

```
1        MOL WT      32.04
         ANTOINE    11.9647 3640.0300  -33.5546
         DENSITY     0.0356    0.0000    0.0000
    L ENTHALPY -0.26477E+06  0.49716E+02  0.11890E+00
    G ENTHALPY -0.21419E+06  0.33037E+02  0.34107E-01


2        MOL WT      18.02
         ANTOINE    11.7927 3887.1001  -43.1718
         DENSITY     0.0222    0.0000    0.0000
    L ENTHALPY -0.30713E+06  0.68215E+02  0.10982E-01
    G ENTHALPY -0.25160E+06  0.30883E+02  0.44509E-02


3        MOL WT      73.10
         ANTOINE     9.7478 3540.8682  -62.7700
         DENSITY     0.0811    0.0000    0.0000
    L ENTHALPY  0.00000E+00  0.00000E+00  0.00000E+00
    G ENTHALPY  0.30000E+05  0.00000E+00  0.00000E+00
```

WILSON COEFFICIENTS:

```
                    -1234.9000 3758.6001
                     4348.3999-4541.7998
                     3592.8999 -763.1000
```

EQUIPMENT

```
BOILER    TYPE 1              H.T.COEFF.   2.0000
  DIA                 1.40
  LENGTH              2.75
  JACKET AREA        12.43
  ADDITIONAL AREA     2.00
  VOLUME              4.23
  HEATING VAPOUR TEMP 423.00


CONDENSER                     H.T.COEFF.   2.0000
  AREA                5.00    HOLD-UP      0.0300
  C.W.TEMPS.,IN/OUT 293.00                303.00


COLUMN
  DIA                 0.40    HEIGHT       3.00
  NO OF PLATES        6       HOLD-UP      7.00
  F-FACTOR            2.50 SPEC PD/PLATE   2.40 MBAR/PLATE
  OPERATING PRESSURE  1.02
```

OPERATION

PRODUCT	STARTING REFLUX	FINAL REFLUX	PURGE REFLUX	HEAVIES SPEC.	LIGHTS SPEC.	PHASE CODE
1	0.50	5.00	4.00	0.0500	0.0000	0
2	0.50	5.00	4.00	0.0250	0.0250	0
BOTTOMS				0.0000	0.0500	

```
INITIAL CHARGE    3.81[M3]  MOLE FRACTIONS  0.500  0.300  0.200

       MAXIMUM TIME    12.00        STEP SIZE   0.10000
```

```
DO YOU WANT NORMAL OR EXTENDED INFORMATION
  [0=NORMAL,1=EXTENDED,]:
1
HOW MANY RECORDS TO PRINT OUT:
1
RELATIVE VOLATILITY IN BOILER   =      3.768


PLATE X1     X2     X3     Y1     Y2     Y3   LIQ.RATE VAP.RATE T     P
  2  0.5993 0.3514 0.0493 0.8650 0.1317 0.0034   9.28   27.83 347.8 1.02
  3  0.6488 0.3411 0.0101 0.8802 0.1191 0.0006   9.28   27.83 346.2 1.02
  4  0.6945 0.3036 0.0019 0.8993 0.1005 0.0001   9.28   27.83 344.9 1.02
  5  0.7519 0.2477 0.0003 0.9228 0.0772 0.0000   9.28   27.83 343.5 1.02
  6  0.8222 0.1777 0.0001 0.9484 0.0516 0.0000   9.28   27.83 341.9 1.02
  7  0.8991 0.1009 0.0000 0.9727 0.0273 0.0000   9.28   27.83 340.2 1.02


FRACTION NR:  1

  TIME   REFLUX  P[BAR] J.TEMP. B.TEMP. B.CONT.    XB      XD    C.LOAD  FLOOD


  0.00    0.50   1.02  362.28  352.28   79.13  0.5000 0.9727
INTEGATION STEP LENGTH      0.2228E-01
CONDENSER AREA REQUIRED:       3.29 M2
F-FACTOR:       1.81 DESIGN F-FACTOR.:      2.50
VAP.FROM BOILER:    27.8257 KMOL/H
 BOILER PRESSURE (BAR)= 1.02      DELTA P= 0.0025
  0.00    0.50   1.02  362.28  352.28   80.00  0.5000 0.9727   0.659   0.725

DO YOU WISH TO CONTINUE UNCHANGED?
DO YOU WANT TO STAY WITH SAME FRACTION?
OUTPUT NORMAL[=0] OR EXTENDED[=1];
0
HOW MANY RECORDS;
10
ENTER NEW TC-GAIN IF DESIRED;

ENTER NEW REFLUX RATIO IF DESIRED;

ENTER NEW RR-GAIN IF DESIRED ;

NEW STEP LENGTH OR TERMINATE THE DISTILLATION
[DEFAULT IS STEP LENGTH;; -1 TO TERMINATE ]

RELATIVE VOLATILITY FOR CALCN; -1; 0; OR ALPHA; :


FRACTION NR:  1

  TIME  REFLUX P[BAR] J.TEMP. B.TEMP. B.CONT.    XB      XD    C.LOAD  FLOOD
  0.02   0.43   1.02  365.26  352.51   79.58  0.4957 0.9600   0.817   0.910
  0.04   0.41   1.02  366.63  352.74   79.17  0.4922 0.9544   0.882   0.989
  0.05   0.40   1.02  366.99  352.94   78.77  0.4891 0.9523   0.887   0.997
  0.07   0.40   1.03  367.21  353.10   74.74  0.4862 0.9502   0.886   0.997
  0.22   0.40   1.03  368.33  354.18   74.37  0.4605 0.9411   0.849   0.963
  0.23   0.43   1.03  368.98  354.30   74.01  0.4580 0.9473   0.883   0.996
  0.24   0.44   1.03  369.16  354.42   73.66  0.4554 0.9470   0.884   0.997
  0.26   0.45   1.03  369.33  354.54   70.14  0.4530 0.9500   0.887   0.997
  0.39   0.45   1.03  370.46  355.62   69.82  0.4276 0.9385   0.851   0.965
  0.40   0.49   1.03  371.08  355.72   66.69  0.4253 0.9485   0.886   0.996

DO YOU WISH TO CONTINUE UNCHANGED?

FRACTION NR:  1

  TIME  REFLUX P[BAR] J.TEMP. B.TEMP. B.CONT.    XB      XD    C.LOAD  FLOOD
  0.52   0.50   1.03  372.26  356.84   66.41  0.4003 0.9373   0.853   0.967
  0.54   0.54   1.03  372.86  356.94   66.13  0.3980 0.9482   0.887   0.997
  0.55   0.55   1.03  373.03  357.06   63.38  0.3957 0.9502   0.889   0.997
  0.66   0.55   1.03  374.23  358.21   63.13  0.3711 0.9372   0.856   0.970
  0.67   0.61   1.03  374.81  358.31   60.67  0.3689 0.9495   0.890   0.997
  0.77   0.61   1.03  376.04  359.49   60.45  0.3450 0.9357   0.858   0.972
  0.78   0.68   1.03  376.61  359.59   58.27  0.3429 0.9512   0.894   0.998
  0.88   0.67   1.03  377.85  360.79   58.07  0.3197 0.9358   0.862   0.973
  0.89   0.75   1.03  378.41  360.89   57.88  0.3177 0.9521   0.897   0.998
  0.89   0.74   1.03  378.57  361.01   55.95  0.3155 0.9491   0.894   0.997
```

DO YOU WISH TO CONTINUE UNCHANGED?

FRACTION NR: 1

TIME	REFLUX	P[BAR]	J.TEMP.	B.TEMP.	B.CONT.	XB	XD	C.LOAD	FLOOD
1.37	1.31	1.03	388.82	368.73	48.24	0.1883	0.9440	0.909	0.998
1.38	1.37	1.03	388.97	368.83	48.14	0.1869	0.9553	0.921	0.999
1.39	1.34	1.03	389.10	368.94	48.05	0.1853	0.9462	0.912	0.998
1.39	1.38	1.03	389.24	369.04	47.96	0.1838	0.9522	0.918	0.999
1.40	1.37	1.03	389.37	369.15	47.87	0.1823	0.9470	0.914	0.998
1.40	1.40	1.03	389.52	369.26	47.77	0.1808	0.9517	0.919	0.999
1.41	1.39	1.03	389.65	369.36	47.68	0.1793	0.9471	0.915	0.998
1.41	1.43	1.03	389.79	369.47	47.59	0.1779	0.9520	0.920	0.999
1.42	1.41	1.03	389.92	369.57	47.51	0.1764	0.9461	0.914	0.998
1.42	1.46	1.03	390.06	369.68	47.42	0.1750	0.9532	0.921	0.999

DO YOU WISH TO CONTINUE UNCHANGED?

FRACTION NR: 1

TIME	REFLUX	P[BAR]	J.TEMP.	B.TEMP.	B.CONT.	XB	XD	C.LOAD	FLOOD
1.43	1.44	1.03	390.19	369.78	47.33	0.1735	0.9460	0.915	0.998
1.44	1.49	1.03	390.33	369.88	47.25	0.1721	0.9530	0.922	0.999
1.44	1.47	1.03	390.46	369.99	47.16	0.1706	0.9469	0.917	0.998
1.45	1.51	1.03	390.60	370.09	46.32	0.1692	0.9512	0.921	0.999
1.50	1.50	1.03	391.73	371.20	46.24	0.1547	0.9176	0.884	0.983
1.51	1.82	1.03	392.16	371.27	46.17	0.1538	0.9738	0.949	1.002
1.51	1.57	1.03	392.25	371.40	46.10	0.1519	0.9264	0.903	0.995
1.52	1.83	1.03	392.42	371.48	46.02	0.1510	0.9706	0.946	1.001
1.52	1.62	1.03	392.52	371.61	45.95	0.1493	0.9306	0.908	0.996
1.53	1.85	1.03	392.68	371.68	45.88	0.1484	0.9690	0.945	1.001

ENTER NEW TC-GAIN IF DESIRED;

ENTER NEW REFLUX RATIO IF DESIRED;

ENTER NEW RR-GAIN IF DESIRED ;
.5
NEW STEP LENGTH OR TERMINATE THE DISTILLATION
[DEFAULT IS STEP LENGTH;; -1 TO TERMINATE]

RELATIVE VOLATILITY FOR CALCN; -1; 0; OR ALPHA; :

FRACTION NR: 1

TIME	REFLUX	P[BAR]	J.TEMP.	B.TEMP.	B.CONT.	XB	XD	C.LOAD	FLOOD
1.53	1.75	1.03	392.78	371.80	45.17	0.1468	0.9514	0.927	0.997
1.58	1.75	1.03	393.84	372.82	45.11	0.1339	0.9135	0.888	0.984
1.59	1.95	1.03	394.24	372.87	45.04	0.1333	0.9542	0.937	1.002
1.59	1.94	1.03	394.31	372.97	44.98	0.1320	0.9474	0.929	0.998
1.60	1.96	1.03	394.44	373.07	44.92	0.1308	0.9480	0.930	0.999
1.60	1.97	1.03	394.56	373.16	44.86	0.1297	0.9475	0.930	0.999
1.61	1.99	1.03	394.68	373.25	44.80	0.1286	0.9481	0.931	0.999
1.61	2.01	1.03	394.80	373.34	44.74	0.1274	0.9476	0.931	0.999
1.61	2.03	1.03	394.92	373.44	44.68	0.1263	0.9480	0.932	0.999
1.62	2.05	1.03	395.03	373.53	44.62	0.1252	0.9473	0.931	0.999

DO YOU WISH TO CONTINUE UNCHANGED?

FRACTION NR: 1

TIME	REFLUX	P[BAR]	J.TEMP.	B.TEMP.	B.CONT.	XB	XD	C.LOAD	FLOOD
1.62	2.07	1.03	395.15	373.62	44.03	0.1241	0.9490	0.933	0.999
1.67	2.08	1.03	396.11	374.56	43.98	0.1129	0.9087	0.894	0.985
1.67	2.35	1.03	396.49	374.60	43.93	0.1124	0.9561	0.947	1.002
1.68	2.32	1.03	396.54	374.71	43.88	0.1111	0.9470	0.936	0.998
1.68	2.35	1.03	396.67	374.79	43.82	0.1101	0.9481	0.938	0.999
1.69	2.37	1.03	396.78	374.88	43.77	0.1091	0.9469	0.937	0.999
1.69	2.40	1.03	396.90	374.97	43.28	0.1081	0.9492	0.940	0.999
1.74	2.41	1.03	397.79	375.84	43.23	0.0981	0.9058	0.898	0.986
1.74	2.74	1.03	398.14	375.87	43.18	0.0977	0.9584	0.956	1.003
1.74	2.68	1.03	398.18	375.98	43.14	0.0965	0.9462	0.941	0.998

TIME	REFLUX	P[BAR]	J.TEMP.	B.TEMP.	B.CONT.	XB	XD	C.LOAD	FLOOD
1.86	3.28	1.03	400.83	378.15	42.01	0.0729	0.9004	0.908	0.987
1.86	3.77	1.03	401.13	378.17	41.98	0.0728	0.9636	0.975	1.004
1.87	3.64	1.03	401.14	378.27	41.94	0.0717	0.9437	0.952	0.997
1.87	3.74	1.03	401.26	378.33	41.63	0.0710	0.9499	0.959	1.000
1.91	3.74	1.03	401.94	379.00	41.60	0.0642	0.8980	0.912	0.988
1.91	4.32	1.03	402.23	379.00	41.57	0.0642	0.9641	0.981	1.005
1.92	4.16	1.03	402.23	379.11	41.55	0.0631	0.9435	0.956	0.997
1.92	4.27	1.03	402.34	379.16	41.27	0.0625	0.9503	0.965	1.000
1.96	4.27	1.03	402.96	379.77	41.25	0.0564	0.8964	0.916	0.988
1.96	4.95	1.03	403.23	379.77	41.22	0.0565	0.9649	0.987	1.005

DO YOU WISH TO CONTINUE UNCHANGED?

FRACTION NR: 1

TIME	REFLUX	P[BAR]	J.TEMP.	B.TEMP.	B.CONT.	XB	XD	C.LOAD	FLOOD
1.97	4.76	1.03	403.22	379.87	41.20	0.0554	0.9426	0.960	0.997
1.97	4.90	1.03	403.33	379.92	40.96	0.0550	0.9514	0.971	1.000
2.00	4.88	1.03	403.88	380.47	40.94	0.0495	0.8927	0.918	0.988
2.01	5.69	1.03	404.14	380.45	40.92	0.0497	0.9663	0.993	1.005

<< CHANGE FRACTION >>>

TIME	REFLUX	P[BAR]	J.TEMP.	B.TEMP.	B.CONT.	XB	XD	C.LOAD	FLOOD
0.00	4.00	1.03	404.11	380.56	40.66	0.0487	0.7685	0.841	0.982
0.03	4.00	1.03	404.88	380.90	40.40	0.0456	0.7315	0.833	0.992
0.07	4.00	1.03	405.43	381.26	40.16	0.0426	0.6948	0.819	0.992
0.10	4.00	1.03	405.98	381.61	39.91	0.0397	0.6557	0.804	0.992
0.13	4.00	1.03	406.52	381.94	39.67	0.0370	0.6190	0.791	0.992
0.16	4.00	1.03	407.05	382.27	39.44	0.0344	0.5826	0.779	0.992

DO YOU WISH TO CONTINUE UNCHANGED?

FRACTION NR: 2

TIME	REFLUX	P[BAR]	J.TEMP.	B.TEMP.	B.CONT.	XB	XD	C.LOAD	FLOOD
0.19	4.00	1.03	407.56	382.58	39.21	0.0319	0.5478	0.769	0.992
0.21	4.00	1.03	408.05	382.88	38.98	0.0297	0.5154	0.759	0.993
0.24	4.00	1.03	408.52	383.17	38.76	0.0276	0.4846	0.750	0.993
0.27	4.00	1.03	408.97	383.44	38.54	0.0256	0.4543	0.742	0.993
0.29	4.00	1.03	409.41	383.70	38.33	0.0238	0.4279	0.735	0.994
0.32	4.00	1.03	409.82	383.95	38.12	0.0221	0.4007	0.727	0.994
0.35	4.00	1.03	410.21	384.19	37.91	0.0206	0.3755	0.720	0.994
0.37	4.00	1.03	410.59	384.42	37.70	0.0192	0.3521	0.714	0.995
0.40	4.00	1.03	410.96	384.64	37.50	0.0178	0.3310	0.709	0.995
0.42	4.00	1.03	411.30	384.86	37.30	0.0166	0.3107	0.703	0.995

DO YOU WISH TO CONTINUE UNCHANGED?

FRACTION NR: 2

TIME	REFLUX	P[BAR]	J.TEMP.	B.TEMP.	B.CONT.	XB	XD	C.LOAD	FLOOD
0.44	4.00	1.03	411.63	385.06	37.10	0.0154	0.2904	0.698	0.995
0.47	4.00	1.03	411.95	385.26	36.90	0.0144	0.2726	0.693	0.996

<< CHANGE FRACTION >>>

TIME	REFLUX	P[BAR]	J.TEMP.	B.TEMP.	B.CONT.	XB	XD	C.LOAD	FLOOD
0.00	0.50	1.03	413.89	387.10	36.00	0.5441	0.9931	0.589	0.950
0.01	0.36	1.03	414.89	386.74	35.80	0.5506	0.9735	0.605	1.006
0.01	0.37	1.03	414.56	386.59	35.60	0.5534	0.9759	0.600	0.995
0.02	0.37	1.03	414.63	386.50	33.57	0.5546	0.9749	0.601	0.998
0.08	0.37	1.03	415.56	387.38	33.39	0.5309	0.9672	0.589	1.002
0.09	0.41	1.03	415.53	387.41	33.21	0.5299	0.9777	0.592	0.992
0.10	0.40	1.03	415.79	387.44	31.40	0.5286	0.9747	0.594	1.001
0.16	0.40	1.03	416.77	388.46	31.24	0.5028	0.9682	0.582	1.005

DO YOU WISH TO CONTINUE UNCHANGED?

FRACTION NR: 3

TIME	REFLUX	P[BAR]	J.TEMP.	B.TEMP.	B.CONT.	XB	XD	C.LOAD	FLOOD
0.16	0.45	1.03	416.73	388.55	31.08	0.5006	0.9777	0.583	0.993
0.17	0.43	1.03	416.99	388.63	29.49	0.4985	0.9750	0.585	1.002
0.22	0.43	1.03	418.04	389.73	29.35	0.4719	0.9656	0.572	1.009
0.23	0.49	1.03	417.90	389.85	29.21	0.4692	0.9797	0.573	0.991
0.23	0.47	1.03	418.24	389.94	29.07	0.4669	0.9735	0.575	1.004
0.24	0.48	1.03	418.23	390.05	28.93	0.4643	0.9770	0.573	0.998
0.24	0.48	1.03	418.38	390.15	28.79	0.4618	0.9735	0.572	1.003
0.25	0.49	1.03	418.42	390.27	28.66	0.4592	0.9759	0.571	0.999
0.25	0.48	1.03	418.56	390.38	28.52	0.4567	0.9734	0.570	1.002
0.26	0.50	1.03	418.62	390.49	27.20	0.4541	0.9758	0.569	0.999

DO YOU WISH TO CONTINUE UNCHANGED?

FRACTION NR: 3

TIME	REFLUX	P[BAR]	J.TEMP.	B.TEMP.	B.CONT.	XB	XD	C.LOAD	FLOOD
0.30	0.49	1.03	419.83	391.68	27.07	0.4275	0.9656	0.559	1.013
0.31	0.56	1.03	419.60	391.82	26.96	0.4245	0.9806	0.558	0.991
0.31	0.53	1.03	419.95	391.92	26.84	0.4222	0.9723	0.559	1.006
0.32	0.55	1.03	419.91	392.04	26.73	0.4195	0.9776	0.558	0.997
0.32	0.54	1.03	420.09	392.16	26.61	0.4171	0.9733	0.557	1.004
0.33	0.56	1.03	420.12	392.28	26.50	0.4145	0.9765	0.556	0.999
0.33	0.55	1.03	420.27	392.39	26.39	0.4120	0.9739	0.555	1.002
0.33	0.56	1.03	420.32	392.52	25.28	0.4094	0.9749	0.553	1.000
0.38	0.56	1.03	421.57	393.77	25.18	0.3833	0.9624	0.543	1.017
0.38	0.66	1.03	421.25	393.93	25.08	0.3803	0.9853	0.543	0.987

DO YOU WISH TO CONTINUE UNCHANGED?

FRACTION NR: 3

TIME	REFLUX	P[BAR]	J.TEMP.	B.TEMP.	B.CONT.	XB	XD	C.LOAD	FLOOD
0.38	0.58	1.03	421.70	394.01	24.98	0.3784	0.9669	0.542	1.012
0.39	0.65	1.03	421.51	394.17	24.89	0.3755	0.9834	0.542	0.991
0.39	0.60	1.03	421.86	394.26	24.79	0.3735	0.9677	0.540	1.011
0.40	0.66	1.03	421.71	394.41	24.70	0.3706	0.9824	0.540	0.992
0.40	0.61	1.03	423.00	394.51	24.60	0.3685	0.9711	0.539	1.008
0.40	0.65	1.03	423.00	394.71	24.51	0.3658	0.9790	0.557	1.032
0.41	0.63	1.04	423.00	395.03	24.42	0.3635	0.9717	0.546	1.023
0.41	0.66	1.04	423.00	395.31	24.33	0.3608	0.9783	0.542	1.008
0.41	0.64	1.05	423.00	395.48	24.24	0.3585	0.9723	0.535	1.006
0.42	0.67	1.05	423.00	395.65	24.15	0.3559	0.9776	0.534	0.997

DO YOU WISH TO CONTINUE UNCHANGED?

FRACTION NR: 3

TIME	REFLUX	P[BAR]	J.TEMP.	B.TEMP.	B.CONT.	XB	XD	C.LOAD	FLOOD
0.42	0.65	1.05	423.00	395.74	24.06	0.3536	0.9728	0.529	0.998
0.43	0.68	1.05	423.00	395.86	23.98	0.3511	0.9782	0.529	0.992
0.43	0.66	1.04	423.00	395.92	23.89	0.3488	0.9714	0.525	0.995
0.43	0.69	1.04	423.00	396.03	23.81	0.3462	0.9794	0.526	0.988
0.44	0.66	1.04	423.00	396.07	23.72	0.3441	0.9715	0.522	0.994
0.44	0.70	1.04	423.00	396.17	23.64	0.3415	0.9788	0.523	0.987
0.44	0.68	1.04	423.00	396.20	23.56	0.3393	0.9707	0.519	0.993
0.45	0.72	1.03	423.00	396.31	23.48	0.3367	0.9809	0.522	0.985
0.45	0.68	1.03	423.00	396.32	23.40	0.3346	0.9698	0.517	0.994
0.45	0.74	1.03	423.00	396.43	23.32	0.3320	0.9812	0.520	0.985

DO YOU WISH TO CONTINUE UNCHANGED?

FRACTION NR: 3

TIME	REFLUX	P[BAR]	J.TEMP.	B.TEMP.	B.CONT.	XB	XD	C.LOAD	FLOOD
0.46	0.69	1.03	423.00	396.44	23.24	0.3299	0.9694	0.514	0.994
0.46	0.75	1.03	423.00	396.55	23.16	0.3273	0.9818	0.518	0.984
0.46	0.70	1.02	423.00	396.55	23.08	0.3253	0.9689	0.512	0.994
0.47	0.77	1.02	423.00	396.67	23.01	0.3226	0.9825	0.516	0.984
0.47	0.71	1.02	423.00	396.67	22.93	0.3207	0.9670	0.509	0.996
0.47	0.79	1.02	423.00	396.80	22.86	0.3179	0.9853	0.515	0.981
0.48	0.70	1.01	423.00	396.77	22.78	0.3162	0.9626	0.506	0.999
0.48	0.82	1.01	423.00	396.94	22.71	0.3133	0.9882	0.513	0.979
0.49	0.69	1.01	423.00	396.87	22.63	0.3118	0.9543	0.501	1.004
0.49	0.87	1.01	423.00	397.11	22.56	0.3085	0.9912	0.512	0.975

DO YOU WISH TO CONTINUE UNCHANGED?
DO YOU WANT TO STAY WITH SAME FRACTION?
OUTPUT NORMAL[=0] OR EXTENDED[=1];

HOW MANY RECORDS;

 ENTER NEW PC-GAIN IF DESIRED ;

ENTER NEW REFLUX RATIO IF DESIRED;

ENTER NEW RR-GAIN IF DESIRED ;
.5

FRACTION NR: 3

TIME	REFLUX	P[BAR]	J.TEMP.	B.TEMP.	B.CONT.	XB	XD	C.LOAD	FLOOD
0.66	1.29	0.90	423.00	400.85	19.43	0.1908	0.9805	0.442	0.975
0.66	1.26	0.89	423.00	400.77	19.39	0.1895	0.9696	0.440	0.987
0.67	1.31	0.89	423.00	400.85	19.36	0.1875	0.9797	0.443	0.982
0.67	1.28	0.89	423.00	400.83	19.32	0.1862	0.9704	0.440	0.991
0.67	1.33	0.89	423.00	400.92	19.28	0.1843	0.9773	0.441	0.986
0.67	1.31	0.88	423.00	400.94	19.25	0.1828	0.9729	0.440	0.990
0.68	1.34	0.88	423.00	401.01	18.89	0.1810	0.9750	0.439	0.989
0.70	1.34	0.88	423.00	402.17	18.86	0.1650	0.9372	0.403	0.973
0.70	1.60	0.87	423.00	402.26	18.83	0.1619	0.9900	0.420	0.949
0.71	1.44	0.87	423.00	401.87	18.79	0.1621	0.9619	0.419	0.985

DO YOU WISH TO CONTINUE UNCHANGED?

FRACTION NR: 3

TIME	REFLUX	P[BAR]	J.TEMP.	B.TEMP.	B.CONT.	XB	XD	C.LOAD	FLOOD
0.71	1.58	0.86	423.00	401.97	18.76	0.1598	0.9850	0.426	0.973
0.71	1.49	0.86	423.00	401.83	18.73	0.1592	0.9658	0.422	0.992
0.71	1.59	0.86	423.00	401.96	18.70	0.1571	0.9829	0.426	0.981
0.72	1.52	0.85	423.00	401.90	18.67	0.1562	0.9669	0.422	0.994
0.72	1.62	0.85	423.00	402.04	18.64	0.1543	0.9831	0.425	0.983
0.72	1.55	0.85	423.00	401.98	18.61	0.1534	0.9665	0.420	0.996
0.72	1.65	0.85	423.00	402.13	18.58	0.1514	0.9831	0.424	0.984
0.73	1.58	0.85	423.00	402.08	18.56	0.1506	0.9657	0.419	0.997
0.73	1.69	0.85	423.00	402.24	18.53	0.1486	0.9838	0.422	0.983
0.73	1.61	0.84	423.00	402.18	18.50	0.1478	0.9661	0.417	0.997

DO YOU WISH TO CONTINUE UNCHANGED?
DO YOU WANT TO STAY WITH SAME FRACTION?
OUTPUT NORMAL[=0] OR EXTENDED[=1];

ENTER NEW PC-GAIN IF DESIRED ;

ENTER NEW REFLUX RATIO IF DESIRED;

ENTER NEW RR-GAIN IF DESIRED ;
.25
NEW STEP LENGTH OR TERMINATE THE DISTILLATION
[DEFAULT IS STEP LENGTH;; -1 TO TERMINATE]

FRACTION NR: 3

TIME	REFLUX	P[BAR]	J.TEMP.	B.TEMP.	B.CONT.	XB	XD	C.LOAD	FLOOD
0.73	1.66	0.84	423.00	402.34	18.47	0.1458	0.9727	0.416	0.989
0.74	1.68	0.84	423.00	402.39	18.44	0.1444	0.9727	0.416	0.990
0.74	1.69	0.84	423.00	402.45	18.42	0.1430	0.9724	0.415	0.990
0.74	1.71	0.84	423.00	402.50	18.39	0.1416	0.9733	0.414	0.990
0.74	1.72	0.84	423.00	402.54	18.36	0.1403	0.9715	0.413	0.991
0.74	1.75	0.83	423.00	402.61	18.34	0.1389	0.9734	0.412	0.990
0.75	1.76	0.83	423.00	402.65	18.31	0.1376	0.9727	0.412	0.991
0.75	1.78	0.83	423.00	402.70	18.29	0.1363	0.9719	0.410	0.991
0.75	1.80	0.83	423.00	402.76	18.26	0.1349	0.9736	0.410	0.990
0.75	1.81	0.83	423.00	402.80	18.24	0.1337	0.9727	0.409	0.991

DO YOU WISH TO CONTINUE UNCHANGED?

FRACTION NR: 3

TIME	REFLUX	P[BAR]	J.TEMP.	B.TEMP.	B.CONT.	XB	XD	C.LOAD	FLOOD
0.76	1.83	0.83	423.00	402.85	18.21	0.1324	0.9717	0.408	0.992
0.76	1.85	0.83	423.00	402.91	18.19	0.1311	0.9732	0.407	0.990
0.76	1.86	0.82	423.00	402.95	18.16	0.1299	0.9720	0.406	0.992
0.76	1.89	0.82	423.00	403.01	18.14	0.1286	0.9735	0.406	0.991
0.77	1.90	0.82	423.00	403.05	18.11	0.1274	0.9722	0.405	0.992
0.77	1.92	0.82	423.00	403.10	18.09	0.1262	0.9723	0.404	0.991
0.77	1.94	0.82	423.00	403.15	18.07	0.1250	0.9722	0.403	0.991
0.77	1.96	0.82	423.00	403.20	18.04	0.1237	0.9734	0.403	0.991
0.77	1.98	0.82	423.00	403.24	18.02	0.1226	0.9718	0.402	0.992
0.78	2.00	0.81	423.00	403.30	18.00	0.1214	0.9731	0.401	0.991

DO YOU WISH TO CONTINUE UNCHANGED?

```
TIME   REFLUX  P[BAR]  J.TEMP.  B.TEMP.  B.CONT.     XB       XD    C.LOAD  FLOOD
0.90    3.75    0.75   423.00   405.77    17.02    0.0659   0.9768   0.361   0.989
0.90    3.74    0.75   423.00   405.70    17.01    0.0658   0.9741   0.362   0.995
0.90    3.75    0.74   423.00   405.72    17.00    0.0653   0.9722   0.361   0.996
0.91    3.79    0.74   423.00   405.77    16.99    0.0646   0.9726   0.360   0.995
0.91    3.83    0.74   423.00   405.80    16.98    0.0640   0.9728   0.360   0.994
0.91    3.87    0.74   423.00   405.83    16.97    0.0633   0.9718   0.359   0.995
0.91    3.92    0.74   423.00   405.88    16.96    0.0626   0.9723   0.359   0.994
0.91    3.96    0.74   423.00   405.90    16.94    0.0620   0.9725   0.358   0.994
0.92    4.00    0.74   423.00   405.93    16.93    0.0614   0.9726   0.358   0.994
0.92    4.04    0.74   423.00   405.96    16.92    0.0607   0.9727   0.357   0.994
```

DO YOU WISH TO CONTINUE UNCHANGED?

FRACTION NR: 3

```
TIME   REFLUX  P[BAR]  J.TEMP.  B.TEMP.  B.CONT.     XB       XD    C.LOAD  FLOOD
0.92    4.08    0.74   423.00   405.99    16.91    0.0601   0.9715   0.356   0.995
0.92    4.14    0.74   423.00   406.03    16.90    0.0594   0.9732   0.356   0.994
0.92    4.17    0.74   423.00   406.05    16.89    0.0589   0.9715   0.356   0.995
0.93    4.23    0.74   423.00   406.09    16.88    0.0582   0.9732   0.355   0.994
0.93    4.26    0.74   423.00   406.10    16.87    0.0576   0.9727   0.355   0.995
0.93    4.30    0.73   423.00   406.13    16.86    0.0571   0.9725   0.354   0.995
0.93    4.34    0.73   423.00   406.16    16.85    0.0564   0.9710   0.353   0.995
0.93    4.41    0.73   423.00   406.21    16.84    0.0558   0.9739   0.353   0.993
0.94    4.43    0.73   423.00   406.21    16.83    0.0553   0.9715   0.353   0.996
0.94    4.50    0.73   423.00   406.26    16.82    0.0546   0.9729   0.352   0.994
```

DO YOU WISH TO CONTINUE UNCHANGED?

FRACTION NR: 3

```
TIME   REFLUX  P[BAR]  J.TEMP.  B.TEMP.  B.CONT.     XB       XD    C.LOAD  FLOOD
0.94    4.54    0.73   423.00   406.27    16.81    0.0541   0.9722   0.352   0.995
0.94    4.59    0.73   423.00   406.31    16.80    0.0535   0.9718   0.351   0.995
0.94    4.65    0.73   423.00   406.34    16.80    0.0529   0.9728   0.351   0.994
0.95    4.69    0.73   423.00   406.36    16.79    0.0524   0.9718   0.350   0.995
0.95    4.75    0.73   423.00   406.39    16.78    0.0518   0.9739   0.350   0.994
0.95    4.78    0.73   423.00   406.40    16.77    0.0514   0.9710   0.350   0.996
0.95    4.85    0.73   423.00   406.45    16.76    0.0507   0.9724   0.349   0.994
0.95    4.91    0.73   423.00   406.47    16.75    0.0502   0.9727   0.349   0.995
0.96    4.95    0.72   423.00   406.49    16.74    0.0497   0.9729   0.349   0.995
0.96    5.00    0.72   423.00   406.51    16.73    0.0492   0.9716   0.348   0.996
```

DO YOU WISH TO CONTINUE UNCHANGED?

FRACTION NR: 3

```
TIME   REFLUX  P[BAR]  J.TEMP.  B.TEMP.  B.CONT.     XB       XD    C.LOAD  FLOOD
0.96    5.07    0.72   423.00   406.55    16.73    0.0486   0.9723   0.347   0.995
```

<< CHANGE FRACTION >>>

```
0.00    4.00    0.72   423.00   408.25    16.63    0.0458   0.8129   0.282   0.964
```

<<<<<< END OF DISTILLATION >>>>>>

                        SUMMARY OF RESULTS:

```
FRACT.*FRACT. *QUANTI-*               MEAN CONC.
NO.    *TIME  *TY     * X1   * X2    * X3    * X4    * X5    * TOTAL HEAT
       *(HOURS)*(KMOLE)*              (MOLE FRACTIONS)       *   (MJ)
*******************************************************************************
WARMUP*  0.1 *       *       *       *       *       *       *  278.2
REFLUX*  0.2 *       *       *       *       *       *       *  337.6
      *      *       *       *       *       *       *       *
1 (0) *  2.0 * 39.49 * 0.9491* 0.0509* 0.0000* 0.0000* 0.0000*  2736.
      *      *       *       *       *       *       *       *
2 (1) *  0.5 *  4.01 * 0.5029* 0.4971* 0.0000* 0.0000* 0.0000*  1378.
      *      *       *       *       *       *       *       *
3 (0) *  1.0 * 20.18 * 0.0349* 0.9405* 0.0247* 0.0000* 0.0000*  4022.
      *      *       *       *       *       *       *       *
4 (1) *  0.0 *  0.09 * 0.0000* 0.8127* 0.1871* 0.0000* 0.0000*  1393.
      *      *       *       *       *       *       *       *
BOILER*      * 16.63 * 0.0000* 0.0640* 0.9360* 0.0000* 0.0000*

TOTAL *  3.8 * 80.41 *                                       *  5415.
```

7. The Algorithms Involved

The major problem of any multicomponent batch distillation design problem is
that of computer time. A rigorous, integrated, multicomponent distillation
with hold-up model is comparatively straightforward to code, but the resulting
computer costs prohibits its use as a design tool for batch distillation equipment which by the very nature of being batch concerns small throughputs and
small equipment.

Speeds have been increased by a factor of about 100 over the rigorous approach
by the following algorithms:

1. By the treatment of the distillation as a ternary with only two
   components reaching the distillate.
   This enables a ternary plate-to-plate calculation to be carried out,
   but with only a single unknown, on which to iterate the convergence.
   The bi-section method is used to locate this unknown; it is extremely
   stable and so very reliable.

   The following diagram shows how the n-component mixture is collapsed
   to two components for the simulation.

Lumping of Components to Simplify Distillation Simulation

Component	Name
1	Light Impurity
2	Product
3	Heavy Key
4	2nd Heavy Key
5	3rd Heavy Key

4 and 5 combined as 2nd heavy key. Ternary plate to plate → Components in Column Head → Components in Product Fraction

2. The integration step length was taken as the time taken to remove a certain
   fraction of the product remaining in the boiler. One tenth was taken as
   standard. If the control loops had not brought both the product specification
   and column loading within a certain tolerance of specification, then only one
   tenth of the defined step length was taken – hence always some progress was
   made.

   This two-level algorithm required 30-100 steps per product fraction. The
   resulting integration was far from smooth but considered adequate for batch
   distillation design purposes.

3. ## Reflux Ratio Control Algorithm

   The basic DDC control algorithm was used at each integration step:

   $$RR_{new} = RR_{old} (1+g_1 f(\Delta S))$$

where g_1 is the controller gain and
   $f(\Delta S)$ is a function of the difference between the achieved product
         composition and the specification.

To achieve a better linearity over a wide range of compositions

   $f(\Delta S)$ was defined as $(\log(1-x_1) - \log(1-x_1^*))$

where x_1 is the achieved product mole fraction for the last iteration,
and x_1^* is the specification.

The reflux control is also used to match the take-off rate with the quantity
of light component for the intermediate fraction case where 2 liquid phases
have been specified. In this case $f(\Delta S)$ is given by:

$$f(\Delta S) = -DELINT \cdot (1+R)/R = \frac{d(DELINT)}{dR}$$

   where DELINT is the fractional difference between the quantity of light
component in the overheads, and the distillate rate.

This expression represents the sensitivity of DELINT to changes in the
reflux ratio. A suitable value of g_1 was found to be 0.5. This can be modified
during the run to improve stability.

4. Boiler Steam (or Column Pressure) Control Algorithm

This employed the DDC control algorithm.

$\Delta T_{new} = \Delta T_{old}(1 + 92(0.90 - \text{fraction loading}))$
(or P)   (or P)

where ΔT is the temperature difference between boiler contents and heating
medium.

5. Hold-Up Treatment

It is not the intention of the program to describe column dynamics (which
requires orders of magnitude more computer time). Hence, hold-up is treated
as a steady state phenomena, with the held-up liquid always being the same
composition as the liquid in the corresponding position in the column. The
corresponding moles of each component are subtracted from the boiler contents
- which results in lower light concentrations, more difficult separations,
higher reflux ratios and larger intermediate fractions than non-hold-up cases.

This comparatively simple approach, however, gave rise to oscillations
because of the interaction of the reflux control action. This was removed by
dampening the hold-up concentration oscillation by taking a running
mean:

$$x'_{new} = \frac{1}{2}(x + x'_{old})$$

where x refers to profile composition, and x´ to hold-up liquid composition. This is generally effective. However, should oscillations occur
with high hold-ups, then attention to the reflux control gain should
break the resonance and stop the oscillations.

8. Graph and Plot Facilities

The BATCH   program produces a file (BATCH.BIN) which after the simulation
can be used to generate plots of x_D, t_B, reflux rates and pressures vs.
time. The plotting software is not supplied with the program.

The Batch program also has the facility to plot the progress of the distillation during the computation. This is excellent for demonstration purposes, but to the intent designer no more information is presented than he can see from inspection of the computed tables.

Graphs and plots can therefore be used with BATCH   if the user provides subroutines equivalent to GRASS, XAXIS, YAXIS and MORSE suitable for his own graphics system. The function of these subroutines is described in the listing.

9. Notes on Program Implementation

The BATCH   program consists of:

- stage 1: Interactively a data file is produced, or an old one
           is modified. This data remains available after the run.

- stage 2: The simulation of the course of the distillation interactively.

Two general purpose EURECHA libraries of subroutines, LIB and LIB2 must be mounted to provide common read and VLE routines.

If data is to be extracted from the CHEMCO physical properties data bank, then the EURECHA CHEMCO subroutines and the archive of data DBIN must be loaded.   If UNIFAC or UNIQUAC is to be used as the VLE model, then CHEMCO must be loaded.

No graphics are supplied with the program, but the statements calling the SEG graphics library have been left in the program but rendered inactive as comment statements. By replacing these calls by the user's equivalent graphics subroutines, then graphical output is obtained during the simulation. Subroutines involved are GRASS, XAXIS, YAXIS and MORSE.

The program has been written in strict ANSI 1966 FORTRAN (apart from DATA statements) and so the coding is transferable between different machines.

File handling statements are 1977 standard, and so for systems other that those those with 1977 FORTRAN it may be necessary to modify these statements. These statements are clearly marked and refer to OPEN and CLOSE.

The program is designed for interactive use, since it is often necessary to modify the gain parameters of the control loops during the simulation. Batch runs will have to be made in a stepwise fashion, modifying the parameters for the next batch run, when the last batch run indicates where difficulties started.

The coding can be converted from interactive to batch operating mode by activating the appropriate data statements in the BLOCK DATA section of LIB.

```
DATA  /IR/5/,  IW/6/,   IPMT/6/,   MODE/0/   activate for batch mode
DATA  /IR/5/,  IW/5/,   IPMT/5/,   MODE/1/   activate for interactive
                                             terminal mode
DATA  IR/5/,   IW/7/,   IPMT/5/,   MODE/2/   activate for interactive
                                             terminal and printer mode.
```

The results file to printer (7) is stored in file BAT.RES for later printing.

The listing should be consulted for further details and the appropriate assignment of files.

E U R E C H A

The European Committee for the Use of Computers in
Chemical Engineering Education

"INTERN" – COLUMN INTERNALS PROGRAM
M A N U A L
———————

(Program Version October 1984)

Program Author : J. Aittamaa
Helsinki TTK, Finland

Manual Author: L.M. Rose
Technisch-Chemisches Labor
E.T.H. Zentrum
CH-8092 Zürich
Switzerland.

CONTENTS   Page

1. INTRODUCTION

    INTERN is a program for calculating the hydrodynamic aspects of column
internals required for column sizing and internals selection.

The entrainment flood point, pressure drop, downcomer   flooding and plate
efficiency for sieve trays, bubble cap and valve trays can be predicted.
Packing internals are characterized by the Packing Factor, from which the
flood point, pressure drop and HETP can be predicted. Correlations for Sulzer
structured packing are also included.

The program operates both in design and simulation mode, enabling the performance of a given tray to be assessed with a given flow rate, or allowing the tray diameter for a given percentage of flooding to be determined for the given flow conditions.

The program is designed primarily for student exercises, so that the interplay between the various plate dimensions and capacity, pressure drop, efficiency and stability can be studied, and an appropriate choice made for particular design exercises.

To effectively carry out exercises in the selection of column internals demands computer evaluation of the proposals because of the large number of alternatives to be evaluated. As well as trying alternative designs for the type of internal, alternative internals should be investigated; performance of the chosen alternative at different points in the column calculated and finally the performance of the internal at reduced rates investigated. Should these different checks show up weaknesses in the proposed design, then the whole procedure should be repeated with a new internal.

The correlations used within the program are mostly based on those reported in Perry (1). In the case of valve trays two alternative correlations are presented so the student can appreciate the errors that are likely to occur with such methods. Further alternative design details are available for investigation since a choice of three weir types are available to show the effect on the final design.

The proposed methods are not as rigorous as the more detailed correlations reported in the literature. The following simplifications have been made:

- liquid gradients across the plate are ignored for pressure drop and weeping calculations

- head loss under the downcomer inlet weir is ignored for the downcomer flooding

- calculations of plate efficiency is correlated only to froth height; plate design, physical properties, mixing and entrainment are not included

- HETP is related only to packing size.

It is hoped that these approximations will be improved in later versions of the program. The program can be used to determine the performance of an existing column by specifying flows, dimensions and internals type, or can be used in the design mode to determine the diameter for a given set of flow conditions and internals type.

The physical properties data required by the program can either be read in as point data or abstracted from the EURECHA 'CHEMCO' data bank.

The program subroutines are so written that they can be used independently from the main program with other distillation programs in order to deliver hydraulic performance data for the user's own engineering program.

As with the other programs in the EURECHA teaching program project, the program can be used in batch, interactive or interactive plus printer modes. All input is free-read, and the coding is written in strict ANSI 1966 FORTRAN to aid interchangeability.

2. THEORY

    The various correlations used for the calculation are given below:

A. Sieve Plates: The pressure drop is determined as the dry tray pressure
                   drop (H_D) plus the height of liquid over the plate (HL)
    times its aeration factor (β).

    HL is the sum of the outlet weir height plus the height of the crest
    over the weir. The height is determined by the weir formula from Perry (1):

    Type 0   uncorrected Francis weir formula
    Type 1   corrected Francis weir formula (Perry, Fig. 18-16)
    Type 2   formula for serrated weirs
    Type 3   formula for circular weirs

    The relative froth density and aeration factors are taken from Perry (1),
    Fig. 18-15.

    The dry plate pressure drop uses the orifice equation with the coefficient
    of discharge, as represented by Fig. 18-14 in Perry for sieve plates. The
    flooding velocity is determined by the Sounders-Brown-correlation, which
    uses liquid entrainment as the criteria for maximum capacity, as reported
    in Perry.

    The weeping velocity is determined from the weeping correlation, given as
    Fig. 18-11 in Perry, which relates liquid head over the tray to dry plate
    pressure drop plus the pressure required for bubble formation.

    The downcomer filling is taken as a pressure balance between the plates,
    excluding the pressure drops due to liquid flowing down the downcomer,
    under the outlet weir and across the plate. Hence % downcomer filling is
    given by the equation

$$\frac{100}{\text{plate spacing}} \times (\text{plate pressure drop} + \text{clear liquid height on tray})$$

    Plate efficiency is calculated by a correlation assuming a linear relation-
    ship between the number of transfer units and froth height, based upon an
    efficiency of 0.7 for a height of 100 mm. This is only appropriate for
    distilling systems, and is intended to demonstrate the relationship between
    pressure drop and plate efficiency. It should not be used for design purposes.

B. Bubble Cap Plates: This corresponds to the sieve tray procedures with the
                        exception of the calculation of dry tray pressure drop
    and the exclusion of the weeping calculation.

C. Valve Trays: (Glitsch-type, VI, Ballast)
                  Pressure drop calculations are made both according to the
    correlation of Glitsch and also Hoppe. Flooding velocity, downcomer loading
    and plate efficiency are calculated as for sieve plates.

D. Random Packing: Raschig rings, Pall rings, Berl and Interlox saddles
                     are defined solely by the packing factor. It is
    preferably to use packing factors measured from actual separation per-
    formance, but in the absence of this information, calculated packing
    factors (a/ε^3) can be used.

The pressure drop is calculated using the correlation of Norton Stoneware, and the flooding velocity from Eckert's correlation published as Fig. 18-39 in Perry (1). HETP is based on the simple relationship:

$$HETP = \frac{\text{nominal dia. of packing (mm)}}{75}$$

and is intended only as an approximate guide.

E. Sulzer-Structured Packing: Correlation for this type of packing for pressure drop has been taken for the Sulzer Mellapak. This packing has an HETP constant over   its operating range of 0.33 m, and a maximum loading corresponding to an F-factor of 3.0 $kg^{1/2} s^{-1} m^{-1/2}$.

REFERENCES

(1) PERRY, R.H. and CHILTON, C.H.:
The Chemical Engineer's Handbook, 5th edition.
McGraw-Hill, 1973

3) USE AS AN INDEPENDENT PROGRAM

Item	No. of Data Items	Description
a) Flows and Conditions		
1	1-80	Title for conditions
2	4	- Liquid molar flow (kmol/h) - Vapour molar flow (kmol/h) - Temperature (K) - Pressure (bar)
b) Physical Properties		
3	1-80	Title for physical system
4	1	Data source flag: 1 = data from bank 2 = data given

If the data is to be extracted from CHEMCO,

5	1	No of components
6	Ncomp	The bank identification numbers for the components
7	Ncomp	Liquid compositions (mol fractions)
8	Ncomp	Vapour compositions (mol fractions)

(These compositions will be normalised, but Ncomp items must always be given, i.e. a binary system requires 2 mole fractions per phase)

Item   No. of Data Items   Description
_____

or, if data is given:

5a   4   Liquid properties:
                                    - molar volume   (m3/kmol)
                                    - mean molecular wt
                                    - viscosity (kg m-1 s-1 =1000 centipoise)
                                    - Surface tension   (n/m = 10 dynes/cm)

6a   3   Vapour properties:
                                    - molar volume   (m3/kmol)
                                    - mean molecular wt
                                    - viscosity (kg m-1 s-1 = 1000 centipoise)

If the data is given then there are no items 7a and 8a.

c) Equipment Specification

9   1-80   Title for equipment

10   1   Mode flag;-   1 for simulation
                                          2 for design

For the case of simulation:

11   1   Column diameter   (m)

or, in the case of design:

11a   1   % flooding at design rates

12   1   flag for equipment type;
                                      1 for sieve plate
                                      2 for bubble cap
                                      3 for valve tray (Glitch correlation)
                                      4 for valve tray (Hoppe correlation)
                                      5 for packing defined by packing factor
                                      6 for SULZER structured packing (MELLAPAK)

Following this selection of equipment type, is a string of data on
one line(or card) which is dependent on the equipment type.

For Type 1 - Sieve plates
   13   5   - % hole area
                            - % active plate area
                            - hole diameter (mm)
                            - plate thickness (mm)
                            - weir height (mm)

For Type 2 - Bubble Cap

   13   5   - % Riser area
                            - % active plate area
                            - Annular/riser areas ratio
                            - slot area/riser areas ratio
                            - height of liquid seal over riser (mm)

Item	No. of Data Items	Description

For Types 3 and 4 - Valve Trays

13	9	- % hole area
		- % max open valve area (% of tray area)
		- % min closed valve area (% of tray area)
		- % active plate area
		- % total valve area
		- density of valve
		- mass of one valve
		- thickness of valve
		- weir height (mm)

For Type 5 - Packing

13	2	- packing factor (1/m)
		(preferably measured values)
		- Specific Surface , a (m)

For Type 6 - SULZER packing - no item 13 data is required.

If a plate is specified (types 1 - 4) Then further weir data is necessary:

14	1	Weir type:
		0 = normal type weir (Francis formula)
		1 = normal type weir
		(corrected Francis formula)
		2 = serrated weir
		3 = circular weir
15	1	If weir is type 0 or 1: give zero
		If weir is type 2 : give serration angle (rad)
		If weir is type 3 : give weir diameter (mm)

d) Program control

16	1	code to execute, modify data or stop.
		-1 = stop
		0 = calculate
		1 = modify conditions (items 1 and 2)
		2 = modify physical properties (items 3 to 8)
		3 = modify equipment (items 9 to 15)
		4 = modify all sections (items 1 to 15)

Item 16 must be entered before the entered data will move to the
calculation stage. It enables data errors to be corrected if the
program is being run interactively and it enables alternative cases
to be evaluated in one program run. After any data modification
item 16 must again be entered to give the instruction to calculate.

**4. USE AS A SUBROUTINE TO USER'S
ENGINEERING PROGRAM**

The engineering program must supply the appropriate data to replace the
input given in table 1. Only that data needs to be provided corresponding to
the equipment type being selected. For example, if only packing is to be used
then no data corresponding to plates or weirs need be supplied. The subroutine

HYDRAU should be called and, providing the correct flags are set, this sub-routine will determine column diameter,   or % flooding and pressure drop information. For details of local names and transfer methods, the listing should be consulted. This contains a detailed description on how the program can be used with users' own engineering programs.

5. NOTES ON PROGRAM IMPLEMENTATION

The program INTERN is supplied as a main program with the necessary sub-routines to function when physical property data is not requested from the data bank.

When physical property data is to be taken from the CHEMCO physical property data bank, then the CHEMCO system, including DBIN, but excluding GEN and GETDAT, must be loaded. Instructions for this are given in the CHEMCO manual. This requires the loading of a library of subroutines (LIB), some of which are already available in the INTERN package for stand-alone use. Some systems will require that one set of these identical subroutines should be deleted. Those associated with the INTERN, and not LIB, should be deleted.

The program has been designed to work in 3 modes:

- BATCH mode   (0)
- INTERACTIVE mode   (1)
- INTERACTIVE + PRINTER   (2)

The usual way of running the program is in batch mode 0. The student has to prepare all the input data and control information necessary to run the program before the running of the program. The program operates on this prepared data without the intervention of the student and provides the result or the error messages at the end of the run. In case of a fatal error occurring during program execution, the program stops with an error message. To correct or modify the input data the student has to revise the data set and run the program again to obtain new results.

The usual input device for the prepared data is a card reader, the output device is usually a printer. However, either could equally well be a disk file.

In mode 1, the student has a terminal with which he keys in the input data and on which he gets back the results. The program works interactively, i.e. keying in the data and return of the results are done via a man-machine dialogue. Correcting the faults; or modifying data can be done immediately. Prompt text and immediate error checks help the student give the input data correctly.

Mode 2 is the same as mode 1 except that, in addition to the terminal, a printer is also available. In this mode results appear on the printer as well as the terminal.

The most comprehensive mode is 2. The other two modes have been included to enable the programs to be used when hardware is limiting.

The MODE is chosen by activating one of three lines in the BLOCK DATA segment:

```
DATA IR/5/, IW/6/,   IPMT/6/,   MODE/Ø/
DATA IR/5/, IW/5/,   IPMT/5/,   MODE/1/
DATA IR/5/, IW/7/,   IPMT/5/,   MODE/2/
```

IR, IW, IPMT are the unit identifier or input, output and prompt peripheries, respectively. The first line belongs to batch mode, the second one to inter-active mode and the third one to interactive + printer mode. From these three lines let the chosen one remain active and transform the other two into comments by writing a C letter in the first position of the lines. To use the program in another mode, change the position of two C letters and compile the BLOCK DATA segment again.

Subject Index